Metallic Glass–Based Nanocomposites

Metallic Glass–Based Nanocomposites

Molecular Dynamics Study of Properties

Edited by
Sumit Sharma

CRC Press
Taylor & Francis Group
Boca Raton London New York

CRC Press is an imprint of the
Taylor & Francis Group, an **Informa** business

CRC Press
Taylor & Francis Group
6000 Broken Sound Parkway NW, Suite 300
Boca Raton, FL 33487-2742

First issued in paperback 2020

© 2020 by Taylor & Francis Group, LLC
CRC Press is an imprint of Taylor & Francis Group, an Informa business

No claim to original U.S. Government works

ISBN-13: 978-0-367-07670-2 (hbk)
ISBN-13: 978-0-367-77660-2 (pbk)

Library of Congress Cataloging-in-Publication Data

Names: Sharma, Sumit, 1985- author.
Title: Metallic glass-based nanocomposites : molecular dynamics study of properties / Sumit Sharma.
Description: Boca Raton : Taylor & Francis Group, LLC, CRC Press is an imprint of Taylor & Francis Group, 2020. | Includes bibliographical references and index.
Identifiers: LCCN 2019017961| ISBN 9780367076702 (acid-free paper) | ISBN 9780429021992 (ebook)
Subjects: LCSH: Nanocomposites (Materials) | Metallic glasses.
Classification: LCC TA418.9.N35 .S478 2020 | DDC 620.1/18--dc23
LC record available at https://lccn.loc.gov/2019017961

Visit the Taylor & Francis Web site at
http://www.taylorandfrancis.com

and the CRC Press Web site at
http://www.crcpress.com

Dedicated to

My daughter & wife,

Dhriti Sharma & Mrs. Rajni Sharma

&

My guru,

Dr. Rakesh Chandra

Contents

Contents

Preface

This book has resulted due to the guidance of my supervisors, Dr. Rakesh Chandra and Dr. Pramod Kumar at Dr B R Ambedkar National Institute of Technology, Jalandhar, India. The editor is highly grateful to Dr. Nupur Prasad and Dr. Raja Sekhar Dondapati for their invaluable contribution. Without their esteemed guidance, support, and motivation, this book might not have materialized. I am very grateful to my parents, who have always motivated me during my life. There was an urgent need for such a book because there is currently no book available which caters to the needs of students and researchers working in the field of molecular dynamics (MD) simulation of metallic glass composites. This book will provide the readers with an overview of the most commonly used tools for MD simulation of metallic glass composites. The editor has been working in the area of MD simulation of composites for the last nine years. Before that, the editor worked in the area of finite element modeling of composites using NISA and MATLAB®.

Biovia Materials Studio, formerly known as Accelrys Materials Studio, is a versatile tool for atomistic modeling. It is being widely used in industry for the simulation of proteins, DNA, and various other biological substances. It is widely used for predicting the properties of nanomaterials, such as carbon nanotubes, polymers, metals, ceramics, and various types of nanocomposites. In this book, the readers will find all the basic steps necessary for simulating any material on Materials Studio. After reading this book, the readers will be able to model their own problems on the above tool for predicting the properties of metallic glass composites.

Chapter 1 deals with an introduction to metallic glasses. Various terms related to metallic glasses are discussed in detail, such as thermogravimetric analysis, transmission electron microscope (TEM), scanning electron microscope, energy dispersive spectroscopy (EDS) or energy dispersive X-ray analysis (EDX), differential thermal analysis (DTA), differential scanning calorimeter (DSC), and X-ray diffraction machine. This is followed by classification of glasses and routes of synthesis of metallic glasses, such as the melt quenching method, the splat quenching method, etc. Then some properties and applications of metallic glasses are discussed.

Chapter 2 gives a detailed explanation of various types of composites, reinforcements, and matrices. In addition, the metallic glass composites are discussed in detail. Chapter 3 elaborates the molecular dynamics methodology for the modeling of metallic glass composites at the nanoscale. The thermal conductivity of metallic glass composites with carbon nanotube and graphene as reinforcements is predicted using MD simulation. Chapter 4 explains various models for calculating the thermal conductivity of metallic glass composites, such as the Maxwell–Garnett model, Hamilton–Crosser model, etc.

In Chapter 5, the damping properties of metallic glass composites are predicted using MD simulations. Normalized storage and loss modulus are predicted for graphene and carbon nanotube reinforced metallic glass composites for varying temperatures. Lastly, Chapter 6 explains the MATLAB® programming of the metallic glass composites.

An attempt has been made here to cover thoroughly all the tools related to MD simulation of metallic glass composites so that the users working in the field of MD can use this book as the text for modeling their own problems as per their requirement. The editor is highly thankful to all the contributors of this book for their excellent work. The editor will also be highly grateful to the potential readers for sending their valuable suggestions, if any, so that this book can be improved further.

Sumit Sharma

Dr B R Ambedkar National Institute of Technology

Jalandhar, India

MATLAB® and Simulink® are registered trademarks of The MathWorks, Inc. For product information, please contact:

The MathWorks, Inc.
3 Apple Hill Drive
Natick, MA 01760-2098 USA
Tel: 508-647-7000
Fax: 508-647-7001
E-mail: info@mathworks.com
Web: www.mathworks.com

Editor

Dr. Sumit Sharma is working as an assistant professor in the Department of Mechanical Engineering at Dr B R Ambedkar National Institute of Technology (NIT), Jalandhar, India. Before joining this institute, he was working as an assistant professor in the School of Mechanical Engineering at Lovely Professional University, Phagwara. He completed his PhD in Composite Materials from NIT Jalandhar in 2015. He did his MTech in Mechanical Engineering, also from NIT Jalandhar, in 2010. He graduated in Mechanical Engineering (with honors) from Kurukshetra University in 2007.

Dr. Sumit Sharma has been working in the area of molecular dynamics simulation of composites for the last eight years. He has more than thirty research articles in reputed journals, such as *Computational Materials Science, Composites Part B, Composite Science and Technology, Journal of Composite Materials, JOM, IEEE,* etc. He is also the reviewer of various journals, such as *Computational Materials Science, Composites Part B, Composite Science and Technology, Computational Condensed Matter, Carbon,* etc. He is a member of the ASTM Society. He has published one other book: *Molecular Dynamics Simulation of Nanocomposites Using BIOVIA Materials Studio, LAMMPS and Gromacs* (Elsevier, August 2019).

His research interests include molecular dynamics, finite element modeling, strength of materials, materials science and engineering, fracture mechanics, mechanical vibrations, and mechanics of composite materials.

Contributors

Rakesh Chandra
Department of Mechanical Engineering
Dr B R Ambedkar National Institute of
   Technology
Jalandhar, India

Raja Sekhar Dondapati
School of Mechanical Engineering
Lovely Professional University
Phagwara, India

Pramod Kumar
Department of Mechanical Engineering
Dr B R Ambedkar National Institute of
   Technology
Jalandhar, India

Nupur Prasad
Division of Research and Development
Lovely Professional University
Phagwara, India

Uday Krishna Ravella
Department of Mechanical Engineering
Madanapalle Institute of Technology &
   Science
Madanapalle, India

Prince Setia
Department of Materials Science &
   Engineering
Indian Institute of Technology
Kanpur, India

Gaurav Sharma
Department of Metallurgical &
   Materials Engineering
Indian Institute of Technology
Roorkee, India

Sumit Sharma
Department of Mechanical Engineering
Dr B R Ambedkar National Institute of
   Technology
Jalandhar, India

Nitin Thakur
Department of Mechanical Engineering
Dr B R Ambedkar National Institute of
   Technology
Jalandhar, India

1 Introduction to Metallic Glasses

Nupur Prasad

CONTENTS

1.1 INTRODUCTION

Chemical composition and molecular arrangement are two important aspects of materials that largely determine the properties of materials. The question is, what makes glass a glass? Is it the chemical composition or the molecular arrangement? The answer to this question is, it is the molecular arrangement that makes a material a glass and not the chemical composition. Chemical compositions help in glass formation. Glasses behave like solids as they retain their shapes at ambient irrespective of the containers they are kept in, which contrasts with liquids. Liquids take the shapes of the containers in which they are kept. At the same time, glasses show irregular arrangement of molecules/atoms in X-ray diffraction studies. Molecules/atoms are irregularly arranged in liquids. Glasses are solids with liquid-like molecular arrangements. They do not have any long-range order of atomic arrangement. The US National Research Council defines glass as an "X-ray amorphous material with a glass transition temperature (Tg)" [1]. The terms *X-ray amorphous* and *Tg* are discussed in the following sections.

1.2   X-RAY AMORPHICITY

A material is called X-ray amorphous when it shows the characteristic amorphous halo in X-ray diffraction analysis. In other words, the absence of characteristic crystalline peaks in X-ray diffraction analysis suggests that the sample is amorphous. Sometimes also referred to as X-ray amorphous materials. To understand this in depth, we need to understand (i) the Bragg's equation and (ii) the X-ray diffraction pattern obtained after running X-ray analysis of any sample.

The crystalline structure and the Bragg's equation:

The word *crystal* is derived from the Greek word *krustallos*. The meaning of *krustallos* is "ice," or any substance looking like ice crystals (see Figure 1.1a), e.g., a rock crystal or a quartz (see Figure 1.1b). Later, with scientific development, it was established through experimental studies that, in crystals, atoms and ions are regularly arranged, forming a motif or pattern that repeats itself in three-dimensions, and generating a long-range order of atoms and ions, which results in the formation of a crystalline structure [1]. The inter-atomic distance in a crystal is normally in angstrom (Å) range. X-rays are electromagnetic radiations with wavelengths in angstrom units $(1 \text{ Å} = 1 \times 10^{-10} \text{ meters})$ [2,3]. The similarity of wavelength of X-ray and inter-atomic distance makes crystals act as diffraction grating for X-radiation [3,4]. When X-rays of a fixed wavelength and at certain incident angles are shone on crystals, the presence of a long-range order of atom/ions in the crystal reflects the impinging X-rays in such a way that they interfere constructively with each other. This results in the formation of signals of high intensity. Constructive interference can occur only when X-rays can reflect in the same phase. The geometry necessary for constructive interference can be seen in Figure 1.2.

When the above X-ray diagram was analyzed using the theorems of mathematics, Bragg landed up with the following Eq. (1.1).

$$n\lambda = 2d \, \text{Sin}\theta \ldots \tag{1.1}$$

where:
  n = an integer (also known as order of reflection)
  λ = wavelength of the X-ray bombarded to the material
  d = interplanar spacing of the crystal
  θ = angle of incident ray with the surface of the sample

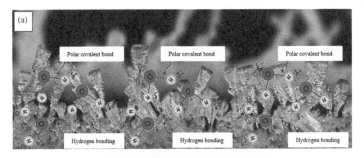

FIGURE 1.1   (a) Image of ice crystals. The overlaying image shows the molecular structure of water molecules. (b) Image of quartz crystals. The overlaying image shows the molecular structure of quartz molecules. (From Prasad, N., and Seddon, A.B., Microwave assisted synthesis of chalcogenide glasses, Thesis, 2010.)

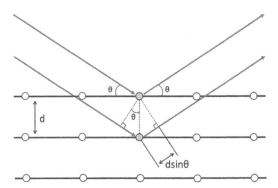

FIGURE 1.2   Path of X-rays with respect to atomic planes of a crystal for the deduction of Bragg's equation (Eq. 1.1). (From Prasad, N., and Seddon, A.B., Microwave assisted synthesis of chalcogenide glasses, Thesis, 2010.)

Hence, because of constructive superimposition of diffracted waves, sharp X-ray diffraction peaks are observed in the case of crystals whereas, when amorphous or glassy materials (where no long-range order of molecules is present) are bombarded with X-rays, the phenomena of superimposition of waves do not occur. This results in the formation of broad diffraction patterns, also known as amorphous hump or amorphous halo. Figure 1.3 shows the difference between the X-ray diffraction patterns

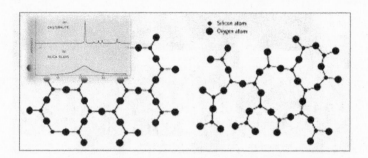

FIGURE 1.3   Powder X-ray diffraction pattern of (a) cristobalite and (b) glassy SiO_2, CuKα radiation. (From Prasad, N., and Seddon, A.B., Microwave assisted synthesis of chalcogenide glasses, Thesis, 2010.)

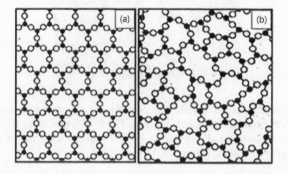

FIGURE 1.4   Schematic molecular arrangement in (a) crystalline (quartz) and (b) amorphous silica. (From Davies, H.A., and Hull J.B., *Nat. Phys. Sci.*, 12, pp. 13–14, 1973.)

of crystalline and amorphous silica. The molecular arrangements of crystalline and amorphous substances are proposed by Zachariasen. The pictorial representations of the molecular arrangements of crystalline and amorphous materials are depicted in Figures 1.4a and b, respectively.

1.3   THE GLASS TRANSITION TEMPERATURE (Tg)

When a glass is heated, it becomes soft at the glass transition temperature (Tg). In Figure 1.5, a material is in the molten state at point "a." The density of the melt increases slowly along the line "ab" as the melt is cooled down. At the liquidus temperature (T_L), the density of most melts increases significantly and suddenly at "bc" in Figure 1.5, leading to the formation of a crystalline solid, whereas some melts behave differently, with no sharp change of density as in the previous case. The change in the density of these melts is along the line "abe," and they form a supercooled liquid. A glass is formed when this supercooled liquid is further cooled to obtain a visco-elastic material having viscosity of the order of 10^{13} Pa s. Tg is the point of inflection of the curve [5–9].

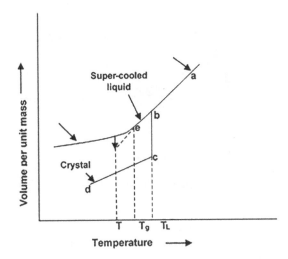

FIGURE 1.5 Volume-temperature diagram showing the difference between a glass-forming material and one that crystallizes on cooling.

1.4 CHARACTERIZATION TECHNIQUES USED FOR GLASS SAMPLE ANALYSIS

A glass is characterized using several advanced materials characterization techniques. These techniques are discussed below.

1.4.1 X-Ray Diffraction Machine

The following are the basic parts of an X-ray diffraction (XRD) machine:

1. *X-ray tube:* This is the source of the X-ray. This normally gives a divergent beam.
2. *Incident beam optics:* This part of the equipment processes the X-ray beam emerging from the X-ray tube before it shines over the sample. To limit the divergence of the incoming beam, it is passed through a Soller slit.
3. *Goniometer:* The word *goniometry* has two terms (i) *gonio* and (ii) *meter.* The term *gonio* is derived from the Greek word *gonia.* The meaning of the word *gonia* is "angle," and the meaning of the term *meter* is "measure." A goniometer is a device which measures an angle. It can also be used to rotate an object at a precise angular position. In an XRD machine, the goniometer holds the sample holder, the incident beam optics, and the detector. It controls the angle between the incident beam and the sample.
4. *Receiving-side optics:* This unit processes the reflected X-ray beam, i.e., after the interaction of the incident beam and the sample.
5. *Detector:* It counts the number of X-rays scattered by the sample.

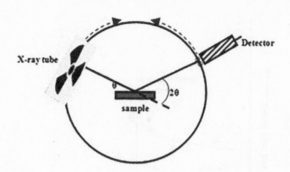

FIGURE 1.6   Path of X-ray in an XRD machine.

Here, θ is the angle between the incident ray and the sample whereas 2θ is the angle between the incident radiation and the diffracted radiation. Since the equipment records the diffracted X-ray counts, they (i.e., the X-ray counts) are plotted against 2θ (see Figure 1.6).

Generation of X-rays: Tungsten filament is heated in a vacuum to generate electrons. These electrons are accelerated toward a fixed target using a high potential field. They decelerate when they hit the target. The process of deceleration leads to the emission of energy in the form of X-rays. A broad, continuous distribution of X-rays called Bremsstrahlung are emitted. X-rays are also generated in one more way, which is discussed in the next few sentences. When electrons impinge a target material, electrons of the inner shells of the atoms (i.e., atoms of the target material) get ejected. Electrons from the outer shells "jump" into these gaps to attain stability. The energy difference between the electron of the inner shell and of the incoming electron is emitted in the form of X-rays. Several materials, viz., copper, chromium, iron, manganese, cobalt, nickel, and molybdenum can be used as target materials [4, 10]. The CuKα radiations are the X-rays obtained when copper is used as the target material [4]. K-alpha emission lines result when an electron falls to the innermost K shell from a 2p orbital of the second or L shell ($n = 2 \rightarrow n = 1$, where n = principal quantum number). The path of X-rays in a typical XRD machine is shown in Figure 1.7a whereas a sample holder with a sample is shown in Figure 1.7b.

The presence of crystalline structures at nanometer scale can be detected using a transmission electron microscope (TEM). Function and construction of a TEM is discussed in the following section.

1.4.2 THE TRANSMISSION ELECTRON MICROSCOPE (TEM)

The function and the construction of equipment of electron microscopes and optical microscopes are similar. Optical microscopes are often referred to as light microscopes. In the optical microscopes (OMs), electromagnetic radiations in visible range are used for investigation. The term "electromagnetic radiations in visible range" signifies the part of it that can be sensed by human eyes, which is within the wavelength region of ~380 nm to ~780 nm, whereas, in electron microscopes (EMs), samples are

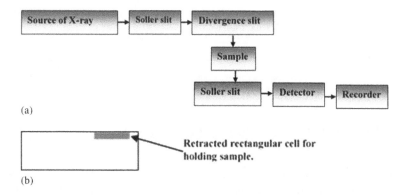

(a)

(b)

FIGURE 1.7 (a) Basic components of an X-ray diffraction machine with path of X-ray. (b) Side elevation of an aluminium sample holder with sample for X-ray diffraction.

exposed to a beam of electrons. Resolution in angstrom units ($1 \text{ Å} = 10^{-10} \text{ m}$) can be achieved via electron microscopes whereas, in optical microscopes, a very small magnification of ~1000 can be achieved. This shows the significance of electron microscopy. It is believed that a researcher working on glass science should have the following basic idea about a TEM and its function.

The equipment: A simple depiction of different parts of OMs (see Figure 1.8a) and EMs (see Figure 1.8b) are shown in the following block diagram.

As depicted in Figures 1.8a and b, both optical microscopes and electron microscopes have roughly the same construction. The difference lies in the source of illuminating agent; for example, in OMs, a sample is illuminated by light whereas, in electron microscopes, a sample is illuminated with electrons. Therefore, a light source is required to use optical microscopes whereas a source of electrons is required to use TEMs. Normally, LaB_6 or a tungsten filament is used for electron generation. In OMs, a glass-condenser lens is required, which condenses the source of light before it hits the sample, whereas, in TEMs, a magnetic counterpart of a condenser lens (normally called a magnetic condenser) is used to attract the electrons coming from the source and focus them toward the sample. Once the light or electron beam (accordingly) hits the sample, the light/electron beam is collected by the objective lens, which forms an image. In light microscopes, the image is seen by the

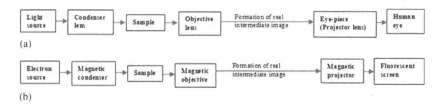

(a)

(b)

FIGURE 1.8 (a) Basic parts of an optical microscope (OM). (b) Basic parts of an electron microscope (TEM).

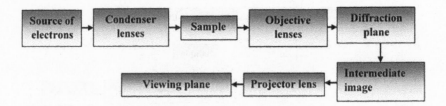

FIGURE 1.9   Schematic of a transmission electron microscope.

human eye whereas, in TEMs, the image is formed on a fluorescent screen. The basic components of TEMs are shown schematically in Figure 1.9. The functions of TEMs are as follows:

- A cathode is a heated filament.
- It is heated using high-voltage electricity.
- A heated cathode produces a ray of electrons that works similar to a beam of light in an optical microscope.
- The condenser lens is constructed of electromagnetic coil.
- The condenser lens concentrates the rays of electrons into a more powerful beam of electrons.
- Another electromagnetic coil focuses the beam of electrons on the central part of the sample.
- The sample is kept on a copper grid in the middle of the main microscope tube.
- The electron beam is passed through the sample, which acquires an image of the sample.
- The projector lens is used to magnify the image. It is visible when the electron beam hits a fluorescent screen, making an image of the sample, which is captured through a computer.

1. *Primary electrons:* The electron beam that emerges from the cathode tube in a TEM is called primary electrons. It is the primary electrons that hit the sample, leading to the occurrence of several phenomena that release electrons and X-rays. These can be captured by keeping the TEM in different modes, which furnishes a lot of important information. In the following paragraph, we discuss these in brief.
2. *Secondary electrons:* When primary electrons come very close to the atoms of the samples, they impart their energy to the electrons of the atoms present in the sample. The electrons of K-shells (normally) gets ejected. These electrons ultimately leave the atoms, forming the secondary electrons. The number of secondary electrons is normally very high as the energy required to eject the electrons present in the K-shells is very low. (It is of the order of ~5 eV.) This results in the formation of bright images when captured for imaging. These electrons are used to study the morphology or the surface structure of samples.

(a)　　　　　　　　(b)

FIGURE 1.10 (a) SAED pattern of a crystalline material. (b) SAED pattern of an amorphous/glassy material.

3. *Selected area electron diffraction:* When a beam of electrons strikes the sample, a large part of it gets scattered. In a sample containing a regular arrangement of atoms (for example, in crystals), the scattered electrons make constructive interference whereas, in materials like glasses (where there is an absence of a long-range order of atoms/ions), there will not be any constructive interference. The underlying principle behind the interference of a diffracted electron beam is similar to that of Bragg's Law (see Section 1.2). When a crystalline substance is studied using a TEM, bright spots demonstrating constructive interference form (see Figure 1.10a) whereas, when an amorphous substance is studied, we get a blurred structure for glassy materials, showing an absence of any long-range order of atoms and hence an absence of constructive interference of the electron beam (see Figure 1.10b).

4. *Auger electrons:* Once the secondary electron is formed in the sample from a K-shell, an electron from the higher shell falls to a lower energy level. This process fills the vacancy created due to the formation of the secondary electron. This leaves the atom in an excited state. The atom can either release its energy in the form of X-rays, or it can release an electron. When an electron is ejected from the sample, it is known as an Auger electron [5–10].

The surface of any glass can be studied using a scanning electron microscope (SEM). The structure and function is discussed below.

1.4.3 Scanning Electron Microscope

A scanning electron microscope (SEM) is used to study the outer surface of any glass, including bulk metallic glasses (BMGs). As the name says, in SEMs a beam of electrons scans the surface of the sample and generates signals. These signals are captured by detectors and displayed by an appropriate display unit, normally a computer screen. As discussed in the section about TEMs, we use an electron microscope to observe any object that cannot be seen by using an optical microscope (OM). To observe any object in nanometers (i.e., a size in the 10^{-9} m range), a scanning electron microscope is usually used.

1.4.3.1 The Equipment

The few important parts of the equipment are discussed in this subsection with different subheadings.

1. *The electron gun:* The job of an electron gun is to produce a beam of electrons. A thin tungsten wire of a diameter less than 1 mm is used as a cathode. The wire is heated to a high temperature of 2800 K, which results in the ejection of a beam of electrons called thermoelectrons. These thermoelectrons are then collected by a positively charged metal anode. A small hole is made at the center of the anode so that the beam of thermoelectrons can flow through the hole. A negatively charged electrode is kept between the cathode and anode to control the size of the beam. The thinnest point of the electron beam is called crossover. Normally, electron beams of 15–20 μm crossovers are used. Lanthanum hexaboride (LaB_6) can also be used as cathodes to produce thermoelectrons.

2. *Condenser lens and objective lens:* To use SEMs, a fine electron beam is required that can be irradiated over the sample. To get the electron beams of the required diameter, magnetic lenses are kept below the electron gun. A magnetic lens is constructed by passing direct electric current through a coil-wound electric wire. The strength of the magnetic lens depends upon the strength of the current. Two magnetic lenses, namely, the condenser lens and the objective lens, are placed under the electron gun. An aperture is constructed of a thin metal plate with a small hole. It is placed between the condenser lens and the objective lens (see Figure 1.11). As the name says, the role of the condenser lens is to condense the beam of electrons, i.e., to narrow down the electron beam. The aperture allows the condensed or narrow beam of electrons to reach the objective lens. The objective lens is used to focus the electron probe to the sample. It also determines the final diameter of the electron probe. These electrons are called primary electrons.

3. *The specimen stage:* The specimen stage of a SEM can move in all three directions. It can be tilted and rotated as well.

4. *Interaction of samples with primary electrons:* As discussed above, the objective lens pushes electrons to the sample. The primary electron beam hits the sample. Several phenomena occur when a primary electron beam interacts with the atoms of the sample. They are as follows:
   Generation of secondary electrons
   Generation of backscattered electrons
   Generation of Auger electrons
   Generation of characteristic X-rays
   Generation of fluorescent X-rays
   Generation of continuum X-rays

The process of interaction of the sample with the primary electrons is pictorially represented in Figure 1.12.

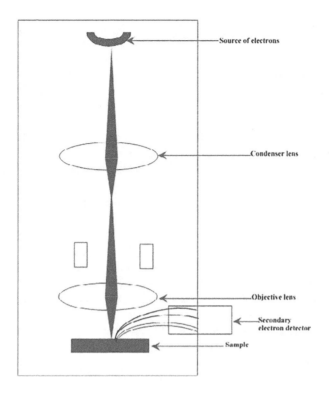

FIGURE 1.11 Schematic of a SEM.

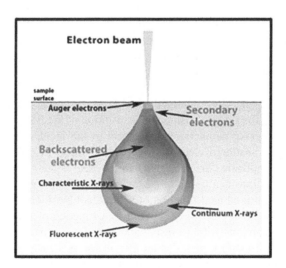

FIGURE 1.12 Schematic showing the occurrence of different phenomena due to the interaction of primary electrons with the sample. (Courtesy of Nanoscience Instruments.)

All these phenomena are described below in brief.

1. *Secondary electron detector:* Secondary electrons are generated when the primary electrons strike the sample inelastically. During this process, electrons are produced from the surface of the sample. These secondary electrons are detected by secondary electron detectors. Secondary electrons are attracted toward the secondary electron detectors because of high voltage. The tip of the secondary electron detector is coated with a fluorescent material. Immense light is produced when secondary electrons fall on the fluorescent tip of the secondary electron detector. This light is then converted into electrons, which generates signals. These signals are then amplified using amplifiers. The amplified signals are then displayed over the display unit as images. These are very useful for surface study and for the inspection of the topography of the sample.

2. *Backscattered electrons (BSE):* These are the electrons obtained as a result of elastic collision of primary electrons with the sample. They are detected by BSE detectors. When primary electrons hit the sample and reach near the nuclei of the sample, an elastic collision occurs. Sometimes these kinds of interactions of electrons with samples are explained by the billiard ball model, in which electrons of small sizes interact with large particles such as the nucleus of an atom. Hence, larger atoms with higher atomic numbers and greater atomic masses produce more BSEs than the atoms of lower atomic number and with lower atomic masses. These phenomena give a direct relation between the number of BSEs detected by the BSE detector and the atomic mass of the sample. The number of BSEs detected by the BSE detector is proportional to the mean atomic number of the sample. When a larger number of BSEs are detected by BSE detectors, a brighter image forms whereas smaller numbers of BSEs detected by BSE detectors result in the formation of less bright or dark images. Hence, in any SEM micrograph, the brighter areas show the presence of heavy atoms whereas the darker area shows the presence of atoms of lower atomic mass.

BSE detectors are generally combined with SEMs. They are normally situated above the sample in the sample chamber. The position of the BSE detector depends upon the BSE scattering geometry with respect to the incident beam. These are solid-state devices. Scattered electrons, i.e., BSEs, travel in different directions; hence, separate components are fabricated in the BSE detectors so that electrons from different directions can be captured for analysis. Detectors above the sample collect electrons scattered as a function of sample composition whereas detectors placed to the side collect electrons scattered as a function of surface topography. BSE images can be obtained almost instantaneously. Using this facility, we can also map our sample. Although it provides an idea of the presence of small and large atoms in the sample, we cannot do compositional mapping using this data. We need to study the characteristic X-rays to get the information about the atoms present in the sample. In the next section, we will see how a SEM provides elemental analysis of the sample.

Energy dispersive spectroscopy (EDS) or energy dispersive X-ray analysis (EDX).
Two types of X-rays are observed when primary electrons strike the sample:

1. *Production of characteristic X-rays:* When primary beam of electrons hit
   the sample, it is possible that electrons present in the atoms of the sample
   collide with the electrons of the primary beam. During inelastic collisions,
   the primary electrons may impart some of their energy to the electrons of the
   atoms of the sample. The electrons of the atoms present in the sample receive
   energy and get kicked out or knocked out of the shell. This process leaves
   a hole or empty shell in the atoms. Electrons present at the higher energy
   state fall to an empty shell of the lower energy state (i.e., to the shell which
   is emptied because of the knocking out by primary electrons). The transition
   of an electron from a higher state to a lower state releases a certain amount
   of energy in the form of characteristic X-rays. This process is shown in the
   diagram in Figure 1.13. The characteristic X-ray produced by each atom is a
   signature of the atom. Analysis of characteristic X-rays provides elemental
   composition of the sample.

   The characteristic X-ray lines are named according to the shell in
   which the initial vacancy occurs and the shell from which an electron
   drops to fill that vacancy. For example, an electron of the K-shell is
   kicked out or knocked out of the atom, creating a hole in the atom in
   the K-shell, and an electron from the adjacent L-shell drops down to the
   empty K-shell. Then the X-ray emitted from this process is named Kα
   radiations. If the electron drops down from the M-shell to the K-shell,
   then the X-ray is named K$_\beta$ radiations. If the atom has a hole in the L-shell
   and the electron drops down from the M-shell, the X-ray is termed Lα-
   radiations. If the electron drops down from the N-shell to the L-shell, the
   radiation is named Lβ radiations. They are known as K-lines, L-lines,
   etc. The nomenclature is depicted pictorially in Figure 1.14.

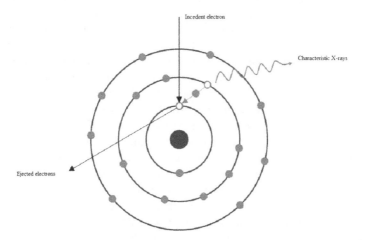

FIGURE 1.13   Schematic showing the production of characteristic X-rays in a SEM.

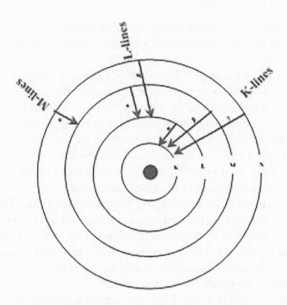

FIGURE 1.14   Schematic showing the nomenclature of characteristic X-ray radiations.

Since K-shells and L-shells are adjacent shells, the probability of transition of electrons from the L-shell to the K-shell is very high. Hence, Kα radiations are one of the strongest lines observed. Kβ-lines are not as probable as Kα-lines, but the energy liberated during transition from the M-shell to the K-shell is higher than that of the energy liberated in the transition of electrons. The data is analyzed using Moseley's Law. A typical data obtained from EDS-analysis is depicted as shown in Figure 1.15.

2. *Continuum (Bremsstrahlung) X-rays:* These are the X-rays generated when the primary-electron beam interacts with the nucleus of the atoms present in the sample. They create a problem for the microscopist as they have to differentiate the characteristic X-ray with that of the continuum X-ray. Continuum X-rays represent the background on which the characteristic X-ray peaks are imposed. They are considered a nuisance by the microscopist because the characteristic X-rays used for elemental identification need to be differentiated from them. The difference in the intensities between continuum X-rays and the characteristic X-rays must be wide enough so that the two can be differentiated and identified. The continuum X-ray must be removed as background before taking the data. The origin of these X-rays is the loss of energy during the interaction between the nucleus of the samples' atoms and primary electrons. The distribution of energy is continuous, and the X-ray produced cannot be used for the elemental analysis.

3. *Auger electrons:* One possible event that may take place during the interaction between primary electrons and the samples' atoms is that it is possible that the excess energy produced during the transition of electrons from higher energy levels to lower energy levels may be consumed by ejecting

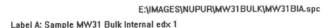

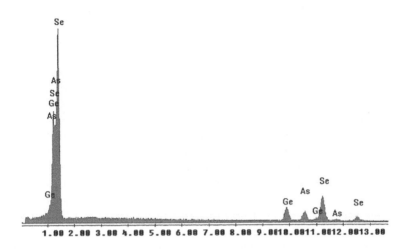

FIGURE 1.15 EDS analysis results of arsenic selenide glass synthesized through microwave heating.

electrons of other atoms (other atoms of the sample). These atoms are called Auger electrons. They have energy in the range of 50–2500 eV. These electrons can be used to study the elemental composition of the sample. They are normally used to identify the light atoms. Its mean free path is 0.1–2 nm and hence are very useful for surface studies.

4. *Fluorescent X-rays:* The highly excited primary electrons, due to interaction with samples, induces X-ray fluorescence as well. This has a very small probe [11–19].

Sample preparation: Normally a very thin layer of the diamond polished sample, in the case of glasses in the form of ingots, is analyzed. Sometimes, the glass ingot is crushed to powder using a clean agate mortar and pestle. The powdered glass is spread over the sticky tab that can be analyzed using a SEM.

The Tg of a glass sample is measured using a differential scanning calorimeter (DSC) and differential thermal analysis (DTA). The working principle and the details of these methods are summarized in the following sections.

1.4.4 DIFFERENTIAL THERMAL ANALYSIS (DTA) AND DIFFERENTIAL SCANNING CALORIMETER (DSC)

These techniques are normally used to determine the glass transition temperature, or Tg value, of any glass sample. These instruments measure the change in temperature of the glass sample with respect to the change in temperature of an inert material, normally called a reference or an inert reference. The glass sample under investigation as well as the inert sample are subjected to a controlled temperature program simultaneously. The change in temperature of the glass sample is recorded in comparison with the inert reference by the respective machines. The temperature of the glass sample as well as the inert sample rises simultaneously until the glass sample does not undergo any physical or chemical change. The occurrence of any change, viz., physical or chemical, for example, decomposition, melting, or a change in crystal structure that either absorbs extra energy or liberates energy, is recorded by these machines. Hereafter, the events that accompany either absorption or liberation of energy happening in the glass sample will be referred to as thermal events. The occurrence of any thermal event is reflected in the form of either (i) the difference in temperature between the inert sample and the glass sample or (ii) the difference in electrical energies required to maintain the same temperature in the glass sample and the inert sample. In DTA, the difference in temperature between the sample and the inert reference during the thermal event is recorded by the net voltage of the back-to-back thermocouples whereas, in DSC, it (i.e., the occurence of the thermal event leading to the difference in temperature between the sample and inert reference) is recorded in the form of electrical power supplied to maintain the equal temperatures of the sample and the inert reference.

A thermal event may be accompanied by the absorption or liberation of energy. When energy is absorbed by the glass sample during the thermal event, the event is called an endothermic process or exothermic reaction whereas, when energy is liberated by the sample, the event is called an exothermic process or exothermic reaction. During an endothermic process in DSC, extra electrical energy is supplied by the system to maintain the equal temperature of the glass sample and the inert sample whereas, during an exothermic reaction, extra energy is supplied to the inert reference to maintain equal temperatures. In DTA, the difference in temperature measured by thermocouples installed back-to-back in the machine tells the story. This is a major difference between DTA and DSC. The difference in the working principle between DTA and DSC shows that, in DSC, enthalpy is directly measured whereas, in DTA, the measurement of change in temperature depends upon the thermal sensitivity of the thermocouple and also at the cost of losing a calorimetric response, demonstrating that DSC gives a more accurate result. Upon the completion of the reactions or thermal event, the temperature of the inert reference and the glass sample will be the same. Mathematically, this process can be understood as follows:

$$\Delta T = T_S - T_R$$

where,

T_S = temperature of the glass sample
T_R = temperature of the inert reference

In DTA, a curve of ΔT is usually plotted against the temperature of the reference. The following Figure 1.16 shows a typical DTA curve of ΔT versus T for an AlF$_3$-based glass obtained by isochronal heating of a small sample of the glass while monitoring the temperature difference (ΔT) of the glass relative to an inert reference supplied with the same rate of heat input. The curve shows the typical endothermic

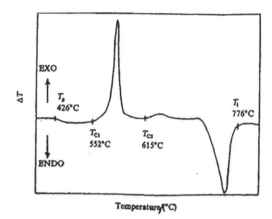

FIGURE 1.16 DTA curve of composition AMCSBY-7 heated at 10°C/min under nitrogen. The sample was in the form of fragments of 0.5–3 mm in size. (A is AlF$_3$, M is MgF$_2$, C is CaF$_2$, B is BaF$_2$, S is SrF$_2$, B is BaF$_2$, Y is YF$_3$.)

change in baseline at Tg as the molar heat capacity of the supercooled liquid >Tg is larger than the molar heat capacity of the glass <Tg. Also, exothermic crystallization peaks Tc_1 and Tc_2 are present on the liquidus when the baseline returns after the melting events. The AlF$_3$-based glass is ideal as it exhibits all these phenomena, giving an exemplar DTA curve. By convention, DTA exothermic peaks are plotted upward and endothermic peaks are plotted downward [9].

The sample preparation: Less than 100 mg. Usually 50–80 mg of the samples are used in the case of DTA. These samples are taken and sealed in a glass tube (see Figure 1.17). Figure 1.18 shows the samples in silica-glass ampoules sitting at the DTA cups ready for Tg determination. The photographs of the sample and reference in the Perkin Elmer DSC Pyris 7 equipment are shown below (see Figures 1.19 and 1.20) [5–9].

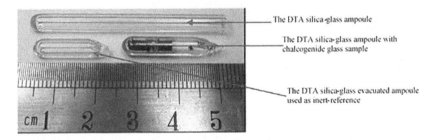

FIGURE 1.17 The DTA silica-glass ampoules ready for the use of determining the glass transition temperature (Tg) of the chalcogenide.

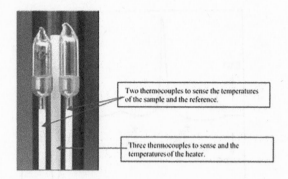

FIGURE 1.18   The DTA silica-glass ampoules containing the chalcogenide glass sample were supported in the DTA cups made of platinum inside the head of the Perkin Elmer DTA 7 Analyzer.

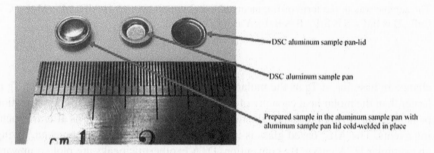

FIGURE 1.19   The DSC aluminium sample pan, sample lid, and the pressed (ready to use) sample lid and pan with the sample inside.

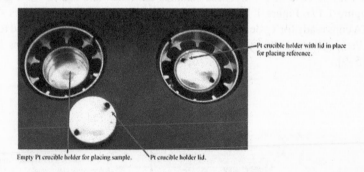

FIGURE 1.20   Plan view of the Perkin Elmer DSC Pyris 7 sample and reference holder.

Bulk metallic glasses undergo several changes when heated. These changes are studied using thermogravimetric analysis. For example, Liaw et al. performed the oxidation test of $Fe_{48}Cr_{15}C_{15}Mo_{14}B_6Er_2$ BMG using TGA [20]. Shek et al. also used TGA for studying the oxidative behavior of $Cu_{42}Zr_{42}Al_8Ag_8$ [21]. We will discuss TGA in this section.

1.4.5 Thermogravimetric Analysis

The term *thermogravimetry* (TG) is composed of two terms: *thermo* and *gravimetry*. The word *thermo* originates from the Greek words *thermos* and *thermē*. The meaning of *thermos* is "hot," and the meaning of *thermē* is "heat." The term *gravimetry* is again composed of two terms: *gravi* and *metry*. The term *gravi* here is the short form of the term *gravity*. The word *gravity* is derived from the Latin word *gravis* or *gravitas*. The meaning of *gravis* is "heavy," and the meaning of *gravitas* is "weight." Now coming to the word *metry*: it is derived from another Greek word, *metria* the meaning of which is "measurer." Combining the meanings of all three terms, we can easily say that there is some relation between heat and measurement of weight. In thermogravimetry, the change in weight of a substance is measured with the change in temperature. We add the term *analysis* at the end to emphasize that the nature of the sample can be studied by analyzing the data obtained in thermogravimetry. The change in weight of a substance is either observed as a function of temperature (also known as scanning mode) or as a function of time (also known as isothermal mode). In scanning mode, heat is fed into the substance by means of an increase in temperature whereas, in isothermal mode, the amount of heat supplied to the substance depends upon time as the temperature of the sample is kept constant. There are several processes (e.g., desorption, absorption, sublimation, vaporization, oxidation, reduction, and decomposition) that depend upon the amount of heat supplied to a substance. Hence, all these processes, named above, can be studied using a TGA machine. One of the important parts of this equipment is the thermobalance. The thermobalance consists of an electronic microbalance, a furnace, a temperature programmer, and a recorder. A block diagram of a microbalance is shown in Figure 1.21 [22, 23].

We have discussed a few of the main characterization techniques used for analyzing BMGs although several other analytical tools are used for this purpose.

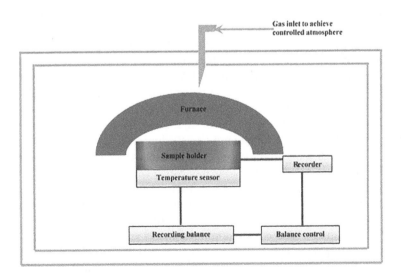

FIGURE 1.21 The schematic of a microbalance.

1.5 TYPES OF GLASSES

The different types of glasses are defined below:

1. *Silica and multicomponent silicate glasses:* Silica (SiO_2) is the main constituent of these glasses. These are based on a Si-O bond. They are normally used for a wide range of commercial applications. We use these in everyday life—e.g., in windows, lamp envelopes, camera lenses, glass containers, etc. These can be either silica or multicomponent silicate glasses. They are transparent from the near-ultraviolet (n-UV, i.e., wavelengths in the range of 200–350 nm) region of EM radiation to the near-infrared (n-IR, i.e., wavelengths in the range of 0.7–2.5 μm) for samples of mm optical path length.

2. *Special glasses:* These glasses are called special glasses, as their transparency of EM radiation is wider than that of silica and multicomponent silicate glasses. It spans from the near-ultraviolet (n-UV, 200 nm) to the mid-infrared (m-IR, >10 μm).

3. *Heavy metal fluoride glasses:* These were fluoride glasses. They were first discovered by Poulain and co-workers in 1974. They discovered heavy metal fluoride glasses based on zirconium and/or hafnium fluoride. These glasses are transparent of EM radiation in the wavelength region of 0.25–6.5 μm, i.e., from the n-UV to the m-IR range, when the optical path length is a few millimeters in thickness. These glasses have low refractive indices. They also show low material dispersion and low linear scattering losses. Hence, these glasses can be used for several applications, e.g., in the fabrication of laser windows in the fabrication of IR sensors, as well as in laser surgery. In summary, it can be said that these glasses can be used for various infrared fiber optics applications. ZBLAN is a widely studied fluoride glass. It is based on six fluoride materials, namely, ZrF_4, BaF_2, HfF_4, LaF_4, AlF_4, and NaF.

4. *Heavy metal oxide glasses:* Dennis and Laubengayer were the first scientists who synthesized GeO_2-based glass. It was synthesized from its melts. A few multicomponent heavy metal oxide glasses, e.g., GeO_2-PbO and TeO_2-ZnO-Na_2O, were also synthesized. These glasses are transparent in the wavelength range from the n-UV (0.4 μm) to the mid-IR (7 μm). The heavy metal oxide glass fibers based on GeO_2 can be used in place of heavy metal fluoride glass fibers for 3-μm laser power delivery applications.

5. *Chalcogenide glasses:* As the name says, these glasses are based on chalcogen elements. S, Se, and Te are classified as chalcogen elements. These glasses, which are based on one or more chalcogen elements in combination with As, Ge, Ga, lanthanides, etc., are called chalcogenide glasses. These glasses are transparent in the range from the mid-visible (0.5 μm) to the mid-IR (~20 μm). Tellurium containing chalcogenide glasses is potentially transparent to wavelengths of more than 15 μm.

6. *Metallic glasses:* These are amorphous metals and their alloys. They are X-ray amorphous, i.e., having no long-range order of atoms/ions with a glass transition temperature (Tg). They possess exceptional properties, e.g., they possess excellent mechanical strength at low temperatures.

They show high flexibility at high temperatures. These glasses show changes in properties at their Tgs that can be exploited for several applications, e.g., shaping. The beauty of metallic glasses lies in the possession of a metallic character with an amorphous structure. Properties, applications, and routes of synthesis of metallic glasses will be discussed in this chapter. Preparation, properties, and applications of metallic glasses in the remainder of the chapter [1,9].

1.6 ROUTES OF SYNTHESIS OF METALLIC GLASSES

The following are the few methods in which metallic glasses were/can be prepared:

1.6.1 THE MELT QUENCHING METHOD

This is the most widely used method for glass synthesis. The three main steps of the melt quenching method are (i) formation of supercooled liquid and (ii) annealing of the supercooled liquid.

1.6.1.1 Formation of Supercooled Liquid

In this process, required amounts of starting materials are taken in a container used for melting (hereafter, it will be referred to as a melt container). The melt container may be in the form of a silica-glass ampoule, a sealed silica-glass ampoule, a silica beaker, or a very unreactive metal crucible, e.g., a platinum crucible, to name a few. The starting materials are heated to prepare a homogenous glass melt. The glass melt is constantly mixed either by melting it in a rocking furnace or by constant stirring. In cases where the viscosity of the melt is very high (~10 poise), it is homogenized by bubbling any suitable gas. For example, to homogenize phosphate glass, oxygen is bubbled throughout the melting process in the glass melt. In another example, chalcogenide glasses are melted in a rocking resistance furnace, which homogenizes the glass melt. In some of the phosphate glass melts, various kinds of stirrers are used for this purpose. In a glass melt, the solid structure of the starting materials breaks. Atoms in the glass melts move freely. In these glass melts, when cooled down rapidly, the viscosity of the glass melts increases abruptly. The process of cooling, in which a large amount of heat energy is extracted from the system (i.e., the glass melt), is called quenching. Quenching brings about the abrupt increase in viscosity. The higher the viscosity of the glass melt, the slower will be the atomic/ionic movement in the glass melt. Quenching leads to solidification. The greater viscosity will pose more hindrance to the mobility of atoms/ions. It takes some time for the atoms to rearrange themselves to form a crystal, but the higher viscosity of the glass melts hinders the ability of atoms/ions to move from the current location to the crystal lattice sites, which is thermodynamically more favorable. This viscous liquid below the freezing point is called supercooled liquid. It is possible to cool the supercooled liquid further with a rise in the viscosity of the liquid without crystallization, leading to the formation of viscoelastic material called glass. The restricted free movement of atoms leads to the solidification of the glass melt. This solid does not have any long-range order of atoms. The process of formation of a disordered system of atoms/ions is further discussed below.

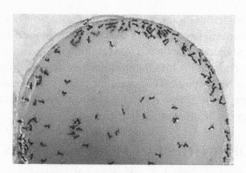

FIGURE 1.22 Photograph of ants dying in honey.

As the temperature falls, molecular motion slows down. If the cooling rate is fast enough, the atoms/ions never reach their equilibrium destination, i.e., thermodynamically favorable crystal lattice points. The substance enters into dynamic arrest, leading to a disordered arrangement of atoms in solid, and a glass forms. These are kinetically stable. Roughly, we can say that during quenching (i.e., the rapid cooling process), there is a tug of war between thermodynamic stability and kinetic stability of the glass melt. A glass forms when kinetic stability wins the war whereas a crystal forms when thermodynamic stability wins the war. The whole process can be brought about in the laboratories as well as in nature because formation of an ordered system takes a certain amount of time since atoms/ions must move from their current location to a more energetically preferred point at crystal nodes, which is enough atoms of the glass melt to solidify and achieve a glassy state. The process can be viewed as the condition of ants in thick honey. Ants cannot move freely in honey (see Figure 1.22), thus losing their more stable life! The quenching process should be such that a suitable amount of heat can be taken out from the melt, taking the melt from a temperature at, or above, the liquidus to below the liquidus. Liquidus temperature is defined as the temperature at which all the solid particles of the starting material get melted to form liquid. The decrease in volume with temperature is then due to the decrease in atomic vibration only. Further cooling of the supercooled melt could produce a viscoelastic material, which, on reaching Tg, i.e., a viscosity of ~$10^{12.5}$ Pas, can be termed as a glass. At these viscosities, the supercooled melt becomes viscoelastic and then freezes to a disordered solid, with mechanical properties often close to an ideal elastic solid.

One smart way of making glasses is changing the stoichiometry of the glass slightly, which hinders the crystal formation. One example is that formation of $GeS_{2.3}$ is easier than the formation of GeS_2.

1.6.1.2 Annealing of the Supercooled Liquid

During quenching, the supercooled liquid, i.e., the solid-like material attains uneven temperature throughout the glass sample. One example of this can be seen in the formation of a contraction cone. A contraction cone forms at the top of the glass sample. It demonstrates that the liquid melt present near the walls or adhered to the walls of the melt container is colder than the liquid present in the interior of the melt during quenching. Consequently, uneven temperature zones lead to uneven viscosities, resulting in the formation of a contraction cone. It is suggested that the uneven temperature is because the glass melt present near to the walls or adhered to the walls of the melt container are more exposed to the cold quenching medium than the glass melt present in the middle, which is surrounded by the immediate neighborhood of hot glass melt. This leads to the creation of uneven viscosity. Hence, the liquid melt present near the walls is at a higher viscosity relative to the material present in the interior of the melt. This creates stress in the glass sample, which can generate poor mechanical properties of the glass samples. The purpose of annealing is to eliminate excess strain in the glass system, which was generated during quenching. This can be done by keeping the supercooled liquid formed after quenching at Tg for a desired period of time. The duration of this time period depends upon the glass composition and its applications. It also depends upon the size of the sample. For example, ~10 g of the supercooled liquid of a chalcogenide-glass sample is kept at Tg for an hour whereas ~5 kg of phosphate glass is kept for ~2 hours at Tg. The sample is then cooled very slowly with an aim that each part of the sample should achieve the same temperature. The slow cooling rate avoids internal stress in the glass-sample.

The concept of glass-forming ability (GFA) is very much discussed in literature. This idea was introduced to understand the factors that are responsible for the formation of glass. GFA is described in brief in the next section.

1.6.1.3 The Glass-Forming Ability (GFA)

GFA can be described as the maximum completely amorphous thickness, or the diameter of any sample produced via melt-quenching method. In general, the GFA of a system increases when the critical cooling rate is higher and decreases with the sample thickness. The amount of heat extracted from the system should be similar in all parts of the sample to make the glass melt viscous enough for vitrification. Apart from this concept, there is Lindsay Greer's confusion principle. According to this principle, a higher number of different-sized species in the glass melts results in a competition for a viable crystal structure. This "atomic confusion" increases the probability of the overall system going into amorphous form. Efforts were made to predict the GFA of metallic-glass compositions. Several models were proposed using computational simulations, but unfortunately satisfactory results could not be obtained. Studies to date show that experimental results obtained are the most reliable ones [24, 25].

1.6.1.4 Melt Quenching Method Used for BMG Formation

In 1960, Duwez et al. synthesized metallic glasses for the first time using the melt quenching method. 25 at% Si-Au mixture was heated to 1300°C to form a glass melt. It was then quenched with a cooling rate of 10^6 K/s to 10^8 K/s when metallic glass was formed, i.e., from 1300°C to room temperature. The samples were very unstable. The authors described in the publication that after 24 hours complex crystals were detected with 11–12 Å interplanar distance [26]. Glassy ribbons of Cu-Zr and Pd-Ge alloy were fabricated using the melt spinning method. The rate of cooling was less than 10^5 °C/sec [27]. Pure Zr can also be turned into a glassy state when cooled with a rate of 10^{14} °C/sec [28].

1.6.2 SPLAT QUENCHING METHOD

In this technique, liquid droplets of the alloy are shot from a carbon dioxide gun onto a cooled Cu or Ag block to produce solid "splats." Amorphous nickel can be produced by this method. Melted nickel (99.99% pure) is cooled down from 1550 °C to room temperature on a water-cooled surface. The critical rate of cooling is between 10^6 °C and 10^{10} °C [29].

1.6.3 PULSED LASER QUENCHING

As discussed above, splat quenching is a type of melt quenching. This is also known as melt spinning. In this technique, a thin layer of glass melt is spread over a cold substrate. The rate of cooling ranges from 10^5 K/s to 10^8 K/s. A very limited number of metallic glasses in a limited composition range can be synthesized using this method. For example, Fe-B-based amorphous alloys can be synthesized using this method but only in the composition range of 12–28 at% of boron (B). A composition having less than 12% of boron resulted in the formation of crystalline (body-centered cubic lattice [bcc]) solids alloys. To overcome these limitations, pulsed-laser facilities are used for rapid melting as well as rapid quenching. Application of laser pulses for extremely short intervals induces a very high cooling rate. For example, a nanosecond pulse laser induces a cooling rate of 10^{10} K/s whereas a picosecond pulse laser brings about a cooling rate of 10^{13} K/s. The achievement of ultrahigh cooling expertise opened a door for the formation of several types of glasses which were not possible otherwise. It was not possible to make such glassy Si with other methods. Similarly, as discussed above, glassy Fe-B alloys having less than 12% boron can also be prepared using pulse laser techniques, although a picosecond pulse laser poses several problems, which are discussed in the following paragraph.

Forming a homogenous glass is a serious problem when a picosecond pulse laser is used for cooling as it is very difficult to homogenize the melt in such a short time period. To address this problem, several research groups used several techniques to homogenize the starting material to that level before melting. There are a few techniques available to solve this problem. For example, ion implantation is a technique that can be used for this purpose. Several research groups used compositionally modulated films with modulation wavelength on the order of a mixing length of 16–35 Å for the formation of glass. A layered structure of Fe and B was created. It was ensured that the defect in the layered structure was not more than 10%. Extra care was taken to remove all the impurities. The Fe-B layered structure was studied using a very sensitive electron probe micro analyzer (EPMA) to ensure that the starting materials contained the least possible impurities. The samples were irradiated with 30-ps neodymium:yttrium aluminium garnet (Nd:YAG) laser pulses (A = 1.06 J-l m). The beam had a Gaussian intensity profile, with a diameter of about l00 J-l m and an average fluence of 0.8 J/cm². Fe-B glasses can be synthesized with this method with a very low amount of boron (5 at%). This success showed a path that glasses of different compositions can be formed using this technique as well [30].

1.6.4 SELECTIVE LASER MELTING

In selective laser melting, metal powders are melted by impinging laser energy on to them. This method has been used for the formation of BMGs. Better results were obtained when higher-energy laser beams were used for this process as high-energy beams helped the melting process. Glasses with high mechanical properties were synthesized using this method. However, this process has a few limitations (e.g., the samples may get cracks as a result of thermal shock). Moreover, initiation of crystallization may occur because of thermal cycling [24].

1.6.5 MEGAPLASTIC DEFORMATION

There are other ways in which metals and their alloys can be amorphized, e.g., mega-plastic deformation (MPD). This involves the forced movement of atoms from their crystal lattice points, which breaks the long-range arrangement. This is also known as amorphization. The glassy state of metal alloys can be obtained via this method. It can be performed at a relatively lower temperature. The deformation of all the samples was brought about by using Bridgman anvils under a quasi-hydrostatic pressure of 4 GPa at room temperature. It is important that this process be thermodynamically favored, i.e., the product is formed only when it is at a lower energy state than the starting materials. There exist at least three factors that determine the tendency toward deformation induced amorphization of crystalline alloys and the corresponding crystalline phases, namely, the mechanical, thermodynamic, and concentration related factors. This phenomenon of solid-state deformation induced amorphization has not yet been studied in detail. In place of a deeper study of solid-state amorphization, it is of great interest and of practical promise to consider an analogy between the tendency to deformation induced amorphization (TDIA) upon MPD and the tendency to thermal amorphization (TTA) upon melt quenching (MQ) [31].

1.6.6 POWDER METALLURGY (SPARK PLASMA SINTERING)

This method solves the problem of sample size where it is difficult to produce samples of larger diameters as well as the challenges involved in fast cooling of the samples. Amorphous powders are produced by different techniques, e.g., mechanical alloying, high-pressure gas atomization, or atomization. The powders are then densified using different procedures, e.g., cold pressing, equal channel angular extrusion, hot pressing, and, the most efficient method, spark plasma sintering (SPS). This method can be improved by the introduction of heat energy during sintering and then rapid cooling. This method has proven to be time-saving and cost-effective, though challenges of incorporation of impurities may pose problems [24].

1.7 PROPERTIES OF METALLIC GLASSES

Properties of materials, e.g., electrical properties and magnetic properties, depend largely on the atomic arrangement of the materials. The long-range periodicity of the atomic arrangement in crystalline solids determines various properties of the crystalline materials. Various geometric lattices of face-centered cubic lattice (fcc) and body-centered cubic lattice (bcc), for example, are the results of long-range atomic arrangement of crystals. Bloch theorem of solid-state physics was proposed based on this conception. Various physical properties, e.g., electrical, magnetic, and mechanical properties, depend largely on these geometric structures. In the following sections, we summarize some of these differences, especially in metallic alloys, and comment on the technological interest in some of the outstanding properties of metallic glasses [24,32].

1.7.1 ELECTRICAL PROPERTIES

Electrical resistivity of crystalline metallic alloys decreases with a decrease in temperature below room temperature. This observation was attributed to a decrease in the scattering of electrons (or holes) by phonons. At lower temperatures, the scattering of electrons (or holes) by phonons decrease, which in turn deceases the overall resistivity of the material. The mean free path of an electron between two scattering events is a function of temperature. It is of the order of a few hundred angstroms at room temperature whereas their values are in the millimeter range at low temperatures, i.e., temperatures lower than room temperature. In contrast to this behavior, the electrical resistivity of amorphous metal alloys is rather small. The estimated value of the mean free path for electrons in metallic glasses is nearly 10 Å. The short mean free path influences the superconducting behavior of the materials.

Superconductors are materials that conduct electricity without resistance below a certain temperature, called critical temperature (Tc). For example, tungsten, mercury, and lead show superconductivity at *ca.* 10 K. A superconductor shows almost zero electrical resistance. These superconducting materials show a change in character when exposed to magnetic fields. They are classified as Type 1 and Type 2 superconductors based on this characteristic. Type 1 superconductors lose their superconductivity abruptly when exposed to a magnetic field with a strength greater than a critical value. This critical value is designated by Hc and is characteristic of a material whereas Type 2 superconductors lose their superconductivity gradually. The amorphous nature of metallic glass influences the superconducting behavior of the material as well. A material showing Type 1 superconductivity in its crystalline state starts showing Type 2 superconductivity after amorphization. The amorphization of materials results in the formation of a short mean free path of electrons, which increases the penetration depth of the magnetic field, leading to the formation of materials that are metallic glasses with Type 2 superconductivity.

Another example of the influence of a short mean free path comes from measurements of extraordinary Hall coefficients in amorphous alloys. The extraordinary Hall coefficients are found to be a factor of 100 larger than in the corresponding crystalline phases. This increase is attributed to an increase in resistivity of the alloys when they have an amorphous structure. It is possible that these large extraordinary Hall coefficients will find applications as sensors for magnetic fields. The addition of small amounts of impurities in the amorphous metallic alloys changes the overall resistive nature of metallic glasses.

1.7.2 Magnetic Properties of BMGs

The property of a material which is responsible for the common phenomena of magnets is called ferromagnetism. It is the ferromagnetism which we observe in everyday life. The existence of ferromagnetic character was first observed in vapor-deposited amorphous films (81 of recent). These films exhibited magnetic bubbles. Magnetic bubbles are tiny, movable, magnetized cylindrical volumes that can be used to store memory bits. Sputtered thin-film alloys of gadolinium and cobalt exhibited magnetic bubbles. This observation attracted attention and efforts are being made to understand the magnetic nature of metallic glasses. Loss of long-range order during the change from a crystalline state to an amorphous state changes the magnetic nature of materials. Materials containing d-block elements are expected to show this change more prominently than materials containing f-block elements as it was believed that band theory is applicable more intensely in the former. The magnetic behavior of materials containing f-block elements is attributed to the electrons present in deep f shells. The long-range order of atomic arrangement of crystalline materials results in the creation of periodic internal electrostatic and magnetic fields whereas, in glasses, each atom has a different local environment. In multi-component systems, the dissimilarity is twofold; those are (i) composition around an atom's (ii) distance from neighboring atoms. Hence, amorphization brings about various changes in the overall magnetic property of a material. Garnet is an excellent example which demonstrates the different magnetic behavior in the amorphous state and the crystalline state.

Curie temperature (T_C), or Curie point, is the temperature above which certain materials lose their permanent magnetic properties. Normally, the Curie temperatures of metallic glasses are lower than that of their crystalline counterparts. It is believed that charge-transfer effects play an important role in this observation. These effects can increase the Curie temperature instead of decreasing it.

Coercivity is the intensity of the applied magnetic field required to reduce the magnetization of a ferromagnetic material to zero. It is a measure of resistance of a ferromagnetic material to become demagnetized. Coercivity is usually measured in oersted or ampere/meter units and is denoted H_C. It can be measured using a magnetometer. Magnetically hard materials have high coercivity whereas soft magnetic materials have low coercivity. BMGs have generally soft magnetic character because of the absence of structural inhomogeneities such as grain boundaries and crystal defects. Grain boundaries and crystal defects hinder the domain wall motion during magnetization. In magnetism, a domain wall is an interface separating magnetic domains. The thickness of the domain wall is 100–150 atoms. A domain wall of soft magnetic material is easily movable. In contrast to this, crystalline metallic alloys show hard magnetic behavior. Very few BMGs, e.g., of Nd-Fe-Al, Pr-Fe-Al and Sm-Fe-Al, show hard magnetic character. These glasses also show the magnetocaloric effect. The magnetocaloric effect is a reversible temperature-changing phenomena that occurs when a material is imposed on a changing magnetic field. This effect can be used for magnetic refrigeration—an energy-efficient and environmentally friendly approach although a few severe limitations were displayed by such BMGs.

1.7.3 Mechanical Properties

As discussed earlier, thermodynamic nonequilibrium, or thermodynamic instability, exists in metallic glasses. To lower the free energy of the system, microscopic structural changes occur. This process is known as structural relaxation. The density of the glass increases with time, which is also known as densification. The coordination state of the atoms also changes. This change alters the atomic transport process severely, which is reflected in the change of properties such as a significant change in density and the alteration of the process of diffusion. The structural relaxation is related to the free volume concept given by Flory and Fox. Free volume is defined as the difference between the average atomic volume and that of dense random packing of hard spheres. It was observed that the density of rapidly quenched glasses is less than less-rapidly quenched glasses, which shows greater free volume and, in turn, suggests lower shear viscosity.

Glasses are annealed to reduce the internal stress. This is reflected in the strikingly different values of tensile strengths of as-quenched glasses and as-annealed glass samples. As-annealed glass samples have five to ten times more tensile strength. The crystalline counterparts of the metallic glasses having the same chemical composition are much more brittle than their glassy counterparts. The tensile response of metallic and polymeric glasses is different. This behavior is attributed to the nature of molecular bonding. In systems having covalent bonds, the changes can occur only when atomic displacements are so large that the inter-atomic bonds are virtually broken. The structure must be cracked locally for any configurational change, and the stresses required for this will be so large that a microcrack, once initiated, propagates throughout the structure. Moreover, metallic glasses show resistance toward corrosion.

1.8 APPLICATIONS OF METALLIC GLASSES

These materials have several unique mechanical properties, e.g., high strength and corrosion resistance. These characteristics make them useful for several unique applications. These applications are discussed below.

1. *Electrocatalysis:* An electrochemical device converts chemical energy into electrical energy. Electron transfer and electrocatalysis are the two main processes that occur in an electrochemical device. A fuel cell is an electrochemical device that changes the chemical energy stored in hydrogen. Similarly, batteries also convert chemical energy into electrical energy. The components of batteries are the same. A battery consists of an anode, a cathode, and a separator. These three components, which are sealed in a vessel with a few terminals and suitable vents, form a battery. A large variety of batteries of different sizes and shapes are available in the market for different purposes, e.g., batteries for mobile phones and laptops, to name a couple. These devices can potentially fulfill the rising demand for energy.

   Electrocatalysts enhance the rate of electrochemical reactions. Catalysts are chemicals/substances that increase the rate of chemical reactions. The relative electrocatalytic properties of a group of materials depend upon the electrical potential at a concentration and temperature. This involves oxidation and reduction reactions through the direct transfer of electrons. Electrocatalysts play an important role in lowering the overpotential of the reactions. Overpotential is the potential difference (voltage) between half reaction's (sometimes referred to as a half-cell reaction) thermodynamically determined reduction potential and the experimentally observed value.

Metallic glasses are promising materials for next-generation energy storage and conversion devices. The absence of a long-range order of molecules results in the formation of a complex electronic structure. BMGs can be used as high-performance electrocatalysts. Their composition and shape can also be changed according to the requirement. They can be used in batteries, microreactors, sensors, and fuel cells. Surface features of BMGs also improve the electrocatalytic process. For example, platinum contains BMGs that are fabricated with unique surface features. These features are fabricated on the surface either by subtractive modification or additive manufacturing. In subtractive modification, atoms that are more reactive in comparison to the atoms of their neighborhood are leached out, which results in the formation of some features (depressions) on the surface of the electrode whereas, in additive manufacturing, new desired substances are deposited on the surface of the electrode via galvanic displacement and underpotential deposition. One major advantage of using BMGs as electrocatalysts is their dealloying property. This process (i.e., dealloying) exposes new active sites. This self-improving property makes them a better candidate for electrocatalysis than conventional Pt/C catalysts. At higher temperatures, they are more stable than conventional Pt/C catalysts. Pt/C catalysts cannot be employed for temperatures greater than 80°C.

Multicomponent electrocatalysts with a large surface area increase the performance of fuel cells. C-nanotubes are normally mixed with Pt to increase the surface area of the catalyst. Various types of nanomaterials, e.g., nanorods, nanotubes, and nanoporous materials, are synthesized for this purpose. Synthesis of all these nanomaterials is difficult and complex. On the other hand, fabrication of multicomponent BMGs is cost-effective and easy. For example, a BMG of Pt-Cu-Ni-P is a wonderful substitute for conventional catalysts as it is easier to fabricate, and its composition can be changed by dealloying. A Pt-rich Pt-Cu-Ni-P BMG shows much better performance than the conventional catalyst. It is also more durable than the conventional Pt/C catalyst.

2. *Application of BMG in water purification:* Efforts are made to use Pt-, Pd-, Fe-, and Ni-based catalyst to degrade AZO dyes and a few organic pollutants. Pt-, Pd-based catalysts are expensive and are susceptible to catalytic poisoning. Fe-, Ni-based catalysts are very reactive and are biotoxic whereas BMGs show exceptional durability and inertness toward chemical reactions. The composition of BMGs can also be changed according to the requirement. For example, an Al-based catalyst can be used for the degradation of AZO-dyes. The crystalline counterpart of the Al-based catalyst does not give

satisfactory results. Shapes of the catalysts remain unchanged. Fe- and Mg-based catalysts can be used for the degradation of organic pollutants.

3. *In the production of hydrogen gas:* Hydrogen is one of the cleanest sources of energy. It is also permanent, independent, and renewable. It can be portable in various phases of matter, e.g., in the form of liquid as water, in the form of gas as hydrogen gas, and in the form of solid as hydrates, such as metal hydrates. It is highly portable and can be transported through pipes. Use of hydrogen as fuel is very environmentally friendly as well (except a little amount of oxides of nitrogen are produced in this process). It can be produced from water, which is in abundance on our Earth. Electrolysis of water is a well-known and cost-effective method that can be used for the production of hydrogen on a large scale. It requires a durable electro-catalyst. Several electrocatalysts, e.g., Pt-based catalysts, metal oxides, car-bides, sulphides, and their combinations with graphitic nanocarbons, have been developed for this purpose. Pt-based catalysts proved to be the most stable as well as the most reactive. A metallic glass made of a mixture of Pd-Ni-Cu-P showed better durability, with a self-stabilizing property. The self-stabilization is perhaps due to dealloying whereas increased activity is due to the absence of a long-range order of atoms in the glass. Thin-film amorphous materials, e.g., a thin amorphous layer of MoS on dealloyed nanoporous gold, forms an excellent electrocatalytic material. The activity is six times better than other MoS-based materials for hydrogen evolution reaction. Large surface area is one of the reasons for better performance.

4. *Biomedical applications:* Traditionally polycrystalline materials are used for the fabrication of biomedical devices. Glasses can be molded at their Tgs with very high precision and hence are/can be used for this purpose. The homogenous and isotropic behavior of BMGs is attributed to the absence of crystalline structures and grain boundaries in the material. Biocompatibility, cytocompatibility, and biocorrosion of this class of mate-rials are widely studied. Zr-, Ti-, Mg-, and Fe-based BMGs showed promis-ing results. BMGs can be used in a wide range of biomedical applications, e.g., orthopedic implantations, for the fabrication of stents (stents are used for cardiovascular diseases), dental fillers and dental implants, surgical blades, pacemakers, medical stapling anvils, and negligibly invasive surgi-cal devices. Ti-based, Zr-based, and Fe-based BMGs are nondegradable and show resistance toward corrosion. They also have exceptional mechanical properties. Hence, they are used in the fabrication of surgical blades, pace-makers, medical stapling anvils, and minimally invasive surgical devices as well as biomedical implants, such as articulating surfaces, artificial prosthe-ses, and dental implants. In contrast to that, Mg-based, Ca-based, Zn-based, and Sr-based BMGs are biodegradable and can be used at places where we need a temporary tool inside the body to repair an organism. For example, they are used in intramedullary needles, bone plates and bone screws, absorbable joints, fillers around dental implants, and fillings of bone after cyst/tumor removal in arthroplasty because they will degrade gradually in the human body after completing their temporary task. The device dissolves

inside the body thus avoiding the necessity of removal of the joining tool, which evades one operation. Mg-based BMGs can also be used for implants. The glassy alloys show better mechanical properties, such as mechanical strength values, low elastic moduli, enhanced corrosion resistance, and flexibility, than their crystalline counterparts. CaMg-Zn BMGs can potentially be used for skeletal applications.

Stent implantation is very widely used for the treatment of cardiovascular diseases. Traditionally, stainless steel (316 L SS) and NiTi are used to make stents. Nickel shows local immune responses in several cases. Besides this, these materials have low mechanical strength and hence a thicker strut is required. In contrast to this, a Zr-based BMG is a promising candidate that can be used in the future for stent fabrication. Its higher mechanical strength, resistance toward biocorrosion, cytocompatibility, and the absence of nickel make it a better material than the traditional materials as thinner stents can be manufactured using Zr-based BMGs. Therapeutics in combination with diagnostics is called theranostics. It can be diagnosis followed by therapy or vice versa. Diagnostics and therapeutics can be codeveloped as well, depending upon the requirements. Glass can be used to fabricate tools for cancer treatment. These devices work at the tumor site only in order to reduce the side effects. Conventionally, ferrous ions are used to produce hydrogen peroxide at the tumor site. Hydrogen peroxide liberates active oxygen, which kills cancer cells. Ferrocene is also used for this purpose. Conventionally used methods have some limitations; for example, overproduction of hydrogen peroxide damages neighboring healthy cells and tissues and ferrous ions are susceptible to oxidation. To overcome these limitations, glassy Fe nanoparticles were used to produce hydrogen peroxide inside the tumor, which removed the cancerous tumor much more effectively than its crystalline counterpart. The flexibility of changing the composition of the glass makes material that can be absorbed inside the human body.

5. *Fabrication of micro fuel cells:* Silicon (Si) is traditionally used for the fabrication of fuel cells. This has several limitations; for example, it has low shock resistance and shows poor electrical conductivity. Si-based micro fuel cells (MFCs) can be made, but the method of fabrication is complex and expensive. In contrast to that, fabrication of BMG-based MFCs is very cost-effective. Lengthwise, there is a wide range of MFCs, from centimeter scale to nanometer scale. The three important components of MFCs, (i) catalyst layer, (ii) diffusion layer, and (iii) flow fields, are made of bulk metallic glasses, using a thermoplastic forming (TPF) process. In TPF, stress-free, small features without pores are formed. At the glass transition temperatures of the BMGs, hot-cutting techniques are used for this purpose. Hot-cutting techniques enable us to fabricate microstructures of BMGs. Pt-based BMGs exhibit outstanding catalytic performance. Zr-based BMGs showed a better performance as current collectors as well as flow fields than conventional counterparts. TPF-based embossing on BMGs is a versatile, time-saving, and cost-effective method to produce novel micro fuel cells.

6. *Current sensor:* The magnetic properties of BMGs are used to create current sensors. A Fe-B-Nd-Nb system that exhibits good fracture toughness, a vast supercooled liquid region, and excellent magnetic properties is used for the fabrication of a current sensor (cantilever).

7. *Wear-resistant gear:* BMGs are proved as very good wear-resistance materials. It was found that gears of centimeter scale made of BMGs furnish much better results than the gears made of conventional materials. The gear-on-gear testing reveals that Cu-Zr-based BMG shows a 60% improvement in wear-resistance properties in comparison to Viscomax C300, which is a very special high-performance steel used by NASA for Mars rover Curiosity.

8. *Microcantilever:* In today's world, a scanning probe microscope is an important device in a laboratory. The microscopes require high-precision microcantilevers. Similarly, nanotribology, catalysis, and magnetic force microscopy are also important areas. These also require microcantilevers. Conventional systems made of silicon, nitrides, and other crystalline materials display certain limitations, e.g., precision in shaping because of grains. Use of BMGs solves these problems.

1.9   SUMMARY

Glasses are the immediate neighbors of crystals. Materials that have a regular atomic arrangement are called crystals whereas materials that do not have a long-range order of atoms are called glasses. Bulk metallic glasses (BMGs) are metallic alloys having no atomic arrangement. The atomic arrangements of these substances are studied using X-ray diffraction (XRD) analysis. Glasses also possess a characteristic glass transition temperature (T_g). It is a range of temperature rather than a specific point. At T_g, the viscosity of the glass melt increases remarkably. This increase in viscosity facilitates the process of formation of glass as at these temperatures the movement of atoms in the glass melt is hindered because of high viscosity, which restricts the ability of the atoms to move to the crystal lattice points thus evading the chance of forming a regular arrangement of atoms. The existence of atoms at the crystal nodes are thermodynamically more favourable, but the hindrance of movement of atoms in the viscous glass melt gives more kinetic stability, which results in the formation of a glass. Hence, a glass is defined as an X-ray amorphous material with a glass transition temperature. Hence, to synthesize a glass, it is required to cool the glass melt in a very short period or instantaneously from a viscous melt to avoid the crystallization process. Bulk metallic glasses (BMGs) are metallic alloys that have no long-range order of atoms and have a glass transition temperature. The biggest challenge of preparation of BMGs is to cool the metal alloy melt quickly. Several methods, e.g., splat quenching and pulsed laser quenching, are employed to attain the required rapid cooling. A few other methods, e.g., megaplastic deformation, selective laser melting, and spark plasma melting, are also used for the synthesis of BMGs. These glasses possess several unique electrical, magnetic, and mechanical properties. Some of these properties exist because of the amorphous nature of the material. The absence of a crystalline structure makes it a better material for shaping. The problem of grain boundary is avoided and hence it can be shaped precisely to nanometer scale. Because of these properties, they can have several applications. They can be used for the fabrication of cardiovascular stents, in electrocatalysis, in water purification, and for hydrogen production, to name a few.

REFERENCES

1. Rawson, H. (1980), *Properties and Applications of Glass (Glass Science and Technology)*, Amsterdam, the Netherlands: Elsevier Scientific Publishing Company, North Holland.
2. https://www.physics.rutgers.edu/grad/506/materials%20crystal%20structure.pdf, Last Accessed 6 November 2018.
3. https://nptel.ac.in/courses/113108054/3, Last Accessed 6 November 2018.
4. http://www.pas.rochester.edu/~blackman/ast104/spectrum.html, Last Accessed 6 November 2018.
5. West, A.R. (1999), *Basic Solid State Chemistry*, 2nd ed. Chichester, UK: John Wiley & Sons.
6. Rawson, H. (1991), *Glasses and Their Applications,* London, UK: Institute of Metals.
7. https://nptel.ac.in/courses/103103026/module2/lec12/2.html, Last Accessed 6 November 2018.
8. Epp, J. (2016), *X-ray Diffraction (XRD) Techniques for Materials Characterization-Materials Characterization Using Nondestructive Evaluation (NDE) Methods*, pp. 81–124. Bremen, Germany: Foundation Institute of Materials Science.
9. Prasad, N. and Seddon, A.B. (2010), Microwave assisted synthesis of chalcogenide glasses, Thesis.
10. http://www.zmb.uzh.ch/static/bio407/assets/Script_AK_2014.pdf, Last Accessed 6 June 2019.
11. https://www.nanoscience.com/techniques/scanning-electron-microscopy/, Last Accessed 24 February 2019.
12. https://serc.carleton.edu/research_education/geochemsheets/techniques/SEM.html, Last Accessed 24 February 2019.
13. https://www.jeol.co.jp/en/applications/pdf/sm/sem_atoz_all pdf, Last Accessed 24 February 2019.
14. https://serc.carleton.edu/research_education/geochemsheets/bse.html, Last Accessed 24 February 2019.
15. https://en.wikipedia.org/wiki/Energy-dispersive_X-ray_spectroscopy, Last Accessed 24 February 2019.
16. http://www.charfac.umn.edu/instruments/eds_on_sem_primer.pdf, Last Accessed 24 February 2019.
17. Prasad, N. (2010), A chalcogenide glass was synthesised via microwave heating. Work performed at University of Nottingham, UK.
18. https://www.xos.com/SEM-XRF, Last Accessed 24 February 2019.
19. https://www.xos.com/SEM-XRF, Last Accessed 6 June 2019.
20. Hsieh, H.H., Kai, W., Huang, R.T., Qiao, D.C., and Liaw, P. K. (2007), "Air oxidation of an $Fe48Cr_{15}C_{15}Mo_{14}B_6Er_2$ bulk metallic glass at 600–725°C," *Materials Transactions*, Vol. 48, No. 7, pp. 1864–1869.
21. Cao, W.H., Zhang, J.L., and Shek, C.H. (2013), "The oxidation behavior of $Cu_{42}Zr_{42}A_{18}Ag_8$ bulk metallic glasses," *Journal of Materials Science*, Vol. 48, pp. 1141–1146.

22. https://www.tainstruments.com/tga-550/, Last Accessed 1 March 2019.

23. https://nptel.ac.in/courses/115103030/23, Last Accessed 3 March 2019.

24. Khan, M.M., Nemati, A., Rahman, Z.U., Shah, U.H., Asgar, H., and Haider, W. (2017), "Recent advancements in bulk metallic glasses and their applications: A review," *Critical Reviews in Solid State and Materials Science*, Vol. 43, No. 3, pp. 233–268.

25. Chen, M. (2011), "A brief overview of bulk metallic glasses," *NPG Asia Materials*, Vol. 3, pp. 82–90.

26. Klement, W.J., Willens, R.H., and Duwez, P. (1960), "Non-crystalline structures in solidified gold-silicon alloys," *Nature*, Vol. 3, pp. 869–870.

27. Budhani, R.C., Goel, T.C., and Chopra, K.L. (1982), "Melt-spinning technique for preparation of metallic glasses," *Bulletin of Materials Science*, Vol. 4, No. 5, pp. 549–561.

28. Zhang, J. and Zhao, Y. (1973), "Formation of zirconium metallic glass," *Nature*, Vol. 430 pp. 332–335.

29. Davies, H.A. and Hull, J.B. (1973), "Amorphous nickel produced by splat quenching," *Nature Physical Science*, Vol. 12, pp. 13–14.

30. Lin, C.J. and Spaepen, F. (1982), "FeB glasses formed by picosecond pulsed laser quenching," *Applied Physics Letters*, Vol. 41, No. 8, pp. 721–723.

31. Glezer, A.M., Sundeev, R.V., and Shalimova, A.V. (2012), "Tendency of metallic crystals to amorphization in the process of severe (megaplastic) deformation," *Doklady Physics*, Vol. 57, No. 11, pp. 435–438.

32. Chaudhari, P. and Turnbull, D. (1978), "Structure and Properties of Metallic Glasses," *Science*, Vol. 199, No. 4324, pp. 11–21.

2 Metallic Glass Nanocomposites
Their Properties and Applications

Nupur Prasad

CONTENTS

2.1 INTRODUCTION

The word *composite* originated from the Latin word *compositus. Compositus* is the past participle of *componere*, the meaning of which is "put together." Similar to the origin of the word, the word *composite* stands for two or more materials combined to form a mixture. You need not have to worry too much about the properties of each constituent of the composite while making a composite material; the only thing you need to keep in mind is that by mixing two or more materials in a composite, you will be able to make a composite which can give you desired properties. Two or more components in different distinguishable phases provide overall synergistic effects to the composite. However, within the composite, one can easily identify different materials as they do not dissolve or blend into each other.

One of the most commonly used man-made composites in everyday life is concrete. Now, the question is, what is concrete? Have you ever been to a building construction site? If not, I strongly recommend you visit those sites before reading this chapter further! You will be able to see concrete. Figure 2.1 illustrates that the mixture of cement, sand, stone chips, and water forms concrete. Concrete can be made manually as well as by concrete mixers [1–3]. These days, a concrete mixer is normally used when large amounts are required for big buildings, for example.

FIGURE 2.1   Process of formation of concrete.

If you see a batch of concrete closely, you can easily differentiate the stone chips present in it. If you touch it, you can feel the coarseness of sand and the softness of cement. It is used in a variety of applications, from buildings to bridges. This is a very useful material that provides a strong foundation and a durable roof for our homes.

Several forms of concrete have been used in the past. It is believed that the time period in which concrete was first invented depends on how one interprets the word concrete. Auburn, Alabama, in the United States created a historical time line of concrete, showing that it was first invented in 12,000,000 BC in Israel. It is stated that spontaneous reactions between limestone and oil shale led to the formation of cement-like materials. Limestone is $CaCO_3$ whereas oil shale is a sedimentary rock containing kerogens. Kerogens are solid mixtures of organic chemical compounds from which liquid hydrocarbons can be produced. Note that there is a difference between kerogen and kerosene. Kerosene, also known as lamp oil, is a flammable hydrocarbon liquid, which is derived from petroleum and potentially can be used as a substitute for conventional crude oil. It has not been used yet as the technique of oil extraction from oil shale is not advanced enough to make it a profitable business. In 3000 BC, Egyptians used concrete made of mud, limestone ($CaCO_3$), and gypsum ($CaSO_4.2H_2O$) for making pyramids. It was found that lightweight aggregate concrete was used in Mohenjo-Daro and Harappa in the Indus Valley civilization.

The Greeks and Romans also used concrete during ancient times. For example, the Roman temple Pantheon was constructed using concrete. Several examples exist all over the world showing the use of concrete in the construction of buildings and canals (e.g., St. Sofia Cathedral of Istanbul, Turkey, and the canal Pont du Gard of France, to name a couple). Different types of concrete were made by mankind during different eras and were used for different purposes [4–6]. Now coming back to composites, nature has also made a few wonderful composites. They are called natural concretes. A brief introduction to natural composites is discussed in the following paragraph.

2.2 NATURAL COMPOSITES

Natural composites are present in animals as well as plants. Wood and bone are natural composites. Wood is formed of cellulose, hemicellulose, lignin, and extractives. Different morphological regions have different concentrations of these components. Cotton contains cellulose but not lignin, which can hold together to give strength. It is the lignin in wood which glues cellulose to form strong wood. The two weak substances—lignin and cellulose—together form a much stronger one.

Wood is a complex material. Have you ever noticed that wood absorbs moisture during the rainy season and releases it during the summer? The process is like breathing moisture. A few scientists raised the question as to whether wood is a living or nonliving thing. Philosophically, at this stage, do we need to redefine the concept of living things or the concept of breathing? We cannot stick to the concept that breathing must involve inhaling oxygen just because we humans breathe oxygen. Wood floats in water because it is filled with pores filled with air. Similarly, bone has an ordered structure having type I collagen fibers reinforced with calcium phosphate. It contains about 25% water. Collagen is also found in hair and fingernails. Collagen on its own cannot be of much use in the skeleton, but in combination with hydroxyapatite, it gives bone all the properties needed to support our body.

Figure 2.2 shows the scanning electron micrographs of cork, illustrating the presence of pores in it. Note that both bone and wood have pores filled with air and both give mechanical strength to the body in which they exist. As discussed earlier, several types of composites are formulated by mankind. The composition of a composite is discussed in the following section (see Section 2.3). Seashells are another example of natural composites. According to Dr. Stephen Eichhorn from the School of Materials at the University of Manchester, "The mechanical properties of shells can rival those of manmade ceramics." The most amazing part is that nature makes it ambient. The formation of pearls inside an oyster is another example [7–10].

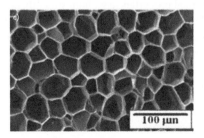

FIGURE 2.2   SEM image of cork. (From SEM image of cork taken by the author (Dr. Nupur Prasad), Brunel University, London, UK.)

2.3   STRUCTURE OF A COMPOSITE

As discussed above, a composite is a combination of two or more materials to form a new material system with enhanced material properties (see Figure 2.3). The matrix and the reinforcement materials are discussed below.

2.3.1   THE MATRIX

A composite contains a continuous phase toughened by a discontinuous phase, also known as a secondary phase. The material used for toughening is known as the rein-forced phase whereas the continuous phase is known as the matrix. The matrix wets the reinforced phase to form bonds so that it can transmit loads onto the reinforced phase. Normally, the matrix of a composite is of lower strength and greater plasticity than the reinforced phase. The greater plasticity of the matrix material allows com-posites to be shaped or molded. There are three basic types of matrices:

1. *Metallic matrix composites (MMC):* As the name says, these composites contain metals as their matrix materials. Several metallic systems (e.g., Al, Be, Mg, Ti, Fe, Ni, Co, and Ag) have been studied and can be used for this purpose. Out of all these, aluminium matrix composites are most extensively used. Ceramics, e.g., SiC, Al_2O_3, B_4C, TiC, TiB_2, and graph-ite, are normally used as reinforcement material. The addition of ceramics provides stiffness, strength, and low density. They can be used to replace cast-iron in engines and brakes. A range of MMCs containing silicon car-bide (SiC), aluminium oxide (Al_2O_3), or other ceramic particles or short fibers in a light alloy, such as aluminium, magnesium, and titanium, can potentially be used for automotive applications. Several such MMCs have been developed that can be used for diesel engine pistons, cylinder liners, brake drums, and brake rotors. Efforts are being made to fabricate MMCs that can be used for the construction of connecting rods, piston pins, and drive shafts. The major impediment toward their wider use is their high cost. Readers can consult Ref. [11] to get a better picture of the importance of composites as NASA is using them for various space applications.
2. *Polymeric matrix:* Composites having polymeric matrices, e.g., thermoplastics or thermosets, are lower in density than MMCs. They have several other unique physical and chemical properties, e.g., high strength, corrosion resistance, and low thermal and electrical conductivity. They also absorb vibrations and do not require surface treatment. Properties of such composites can be controlled by changing the polymer. These materials are used in aircraft design. The

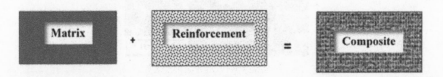

FIGURE 2.3   Schematic showing the general formula of any composite.

Council of Scientific and Industrial Research-National Aerospace Laboratory (CSIR-NAL, India) has constructed civil passenger aircraft named SARAS. The wing and tail of a SARAS aircraft is made of carbon-epoxy composite. Other parts, e.g., spars, ribs, and stringers, of the SARAS aircraft are also made of composite materials in order to reduce the weight of the aircraft. In Kerala (India), a houseboat was designed for tourism. The design was based on traditional boats used for the transportation of rice. Several parts of the traditional houseboat made of wood and steel were replaced by lightweight glass-reinforced polymers, which led to the formation of a modern lightweight houseboat. A detailed discussion of these can be found in Ref. [12–17].

3. *Ceramic and glass matrices:* Ceramics are inorganic and nonmetallic crystalline materials. They are a ubiquitous material. Tile, bricks, plates, glass, and toilets are made of ceramics. Ceramics can be found in products like watches (quartz tuning forks—the timekeeping devices in watches), snow skies (piezoelectric ceramics that stress when a voltage is applied to them), automobiles (sparkplugs and ceramic engine parts found in race cars), and phone lines. They can also be found in space shuttles, appliances (enamel coatings), and airplanes (nose cones). Ceramics of desired properties can be fabricated. Glass is an amorphous substance which is formed by the solidification of the melt without crystallization. The main disadvantages are the considerable brittleness, poor workability, and high sensitivity to internal defects. This aspect of composites is discussed in Ref. [18] in detail.

2.3.2 THE REINFORCEMENT PHASE

The reinforcement phase offers greater strength, modulus of elasticity, and is less prone to deformation than the matrix phase. The size, shape, density, and orientation of this phase determines the overall property of the composite. For example, powder used as the reinforcement phase behaves like the matrix, making the system isotropic, whereas cylindrical reinforcement materials, e.g., fibers as reinforcement material, make the composite anisotropic. They are normally classified according to their geometry as follows:

1. *Particulate reinforcement phase:* Particulate reinforcement is mainly used to improve heat resistance, electrical conductivity, damping of vibrations, wear resistance, hardness, and resistance to high temperatures of the matrix material. It is not very useful for improving the mechanical strength. Particles of any shape (e.g., spherical, pyramidal, and lamellar) can be used for this purpose. Particles of a wide range in sizes, starting from nanometer scale to a few millimeters, are considered particulate reinforcement phase: inorganic powders, e.g., oxides (e.g., MgO, ZnO, BeO, Al_2O_3, ZrO_2), carbides (e.g., SiC, TiC, B_4C, Al_4C_3), nitrides (e.g., Si_3N_4, BN), borides, or silicates (e.g., kaolin, mica, glass beads).

2. *Dispersion-reinforced composites:* In this kind of composite, 10%–15% of the particles, with sizes in the range of 10–100 nm, are used as reinforcing agents. Reinforcement improves mechanical properties like the ultimate

tensile stress and yield stress of the matrix material. It also suppresses the tendency creep of the matrix. Sometimes the use of dispersion reinforcement is favored over precipitate reinforcement as precipitates sometimes tend to melt or coagulate at higher temperatures. Particles of large sizes bear loads, which is a big advantage. The use of porous particles reduces the overall weight of the concrete. It can also be used for various thermal applications as upon evacuation they can show low thermal conductivity behavior.

Various automotive applications, e.g., body panels, bumpers, dashboards, and intake manifolds, are fabricated from plastics reinforced with glass. Brakes are made of particulate composites composed of carbon or ceramic particulates. Toys, electrical products, computer housings, cell phone casings, office furniture, and helmets are made from particulate-reinforced plastics.

Concrete is a particulate composite. The term "concrete jungle" is often used to refer to a city, or an area of a city, with many tower blocks or tenement flats made of brick and concrete. We have replaced green, natural jungles with concrete jungles! Our lives are almost impossible without concrete jungles, and at the same time, life is impossible without green, natural jungles. We must think of a way out! Are we going to preserve our green jungles? Are we going to preserve our concrete jungles from destruction, without which our lives are impossible? Are we going to reuse our used concrete? There are several questions and several problems. As it is said, the best way to escape from a problem is to solve it! We need to think about and solve the problem. The beauty of the human mind is that we neither hesitate in creating a problem nor hesitate in solving it! So, keep your fingers crossed and hope for the best. Our coming generation will create a few more environmentally friendly concretes and will preserve Mother Nature to save Mother Earth.

3. *Fiber reinforcement:* Composites can be reinforced with fibers as well. The strength of fibers depends upon their diameter or cross-sectional area, and it increases with a decrease in the diameter of the fiber. Fibers can be classified in several ways. Table 2.1 depicts the classification of fibers based on diameter. The classification of fibers according to the materials the fibers are composed of is listed in Table 2.2. Fiber-reinforced composites are used in a wide range of applications. For example, the use of glass-fiber-reinforced

TABLE 2.1
Classification of Fibers Based on the Diameter

Fiber Category	Diameter of the Fiber
Nanofibers	Up to 100 nm
Microfibers	0.1–1 µm – whiskers
Middle-sized	1–10 µm
Coarse (rough) fibers	More than 10 µm

TABLE 2.2

Classification Fibers Based on the Constituent Materials of the Fibers

Fiber Category	Description
Natural	Biodegradable and normally composed of cellulose. Examples are as follows: Bast fibers: flax, hemp, jute, kenaf, wood core surrounded by stem containing cellulose filaments Leaf fibers: sisal, banana, palm Seed fibers: cotton, coconut (coir), kapok
Glass	Normally, used to reinforce polymer matrices normally classified as glass-fiber-reinforced plastic (GRP), sometimes also referred to as fiberglass.
Graphite and carbon	Carbon/graphite containing composite Amorphous carbon used for reinforcement Crystalline graphite used for reinforcement
Polymer	Kevlar is a representative example of this class. Polymers require surface treatment because of low wettability.
Ceramic	Several oxides, e.g., MgO, ZnO_2, TiO_2, Al_2O_3, SiO_2, and mixed oxides, e.g., mullite $3Al_2O_3 \cdot 2SiO_2$, or spinel $MgO \cdot Al_2O_3$, carbides (SiC, TiC, B_4C), nitrides, or metal compounds are used for reinforcement of the composite.
Metal	Tungsten/molybdenum are used to reinforce metal alloys for temperature resistance. Steel is used to reinforce aluminium.
Whiskers	Whiskers are single crystals with minimum defects. SiC whiskers are used to reinforce polycrystalline ceramics to fracture toughness.

polymer (GFRP) in the construction of boats is cost-effective and easy to maintain. More than 80% of the exterior of the bodies used in a marine environment are constructed of GFRP. A naval mine is a self-contained explosive device placed in water to damage or destroy surface ships or submarines. Minesweepers are specially designed ships used to remove explosives. GFRPs are used for this application as minesweepers require nonmagnetic materials. They have unique electrical and thermal properties.

Several fabrics are also reinforced that can be used for specific purposes. The process of reinforcing can be done only at defined temperature and pressure conditions. This process is called prepreg. Glass-phenolic resin prepregs are used to improve protection for armored vehicles. Carbon-fiber-reinforced composites (CFRPs) are expensive and hence are used for only specific purposes. CFRPs are used in the construction of aircraft brakes. Aircraft brakes require materials that do not fail at higher temperatures. The usual construction is based on multiple rotating and stationary discs, which can reach surface temperatures of up to 3000 K. The disc material must therefore have excellent thermal and shock resistance and high-temperature strength, together with good thermal conductivity. CFRPs are the ideal materials for this purpose. Apart from this, CFRPs are also used in the fabrication of high-quality sports goods, e.g., tennis racquet frames, golf clubs, and fishing rods. They are also used in the fabrication of laptops

FIGURE 2.4 Molecular structure of Kevlar. The overlaying molecular structure of Kevlar reinforces the image. Incorporation of Kevlar in the manufacturing of safety equipment increases the mechanical properties of the equipment.

and cameras. Race cars, luxury cars, e.g., part of the Mercedes Benz range and the roof of the GM Corvette ZR1, are made of CFRPs.

4. Aramids are aromatic amides. The name is a portmanteau (which is a word made of a combination of two or more words as they sound, to form a single word) of *aromatic polyamide*. In this fashion, the words *aromatic* and *polyamide* are blended together to form the word *aramids*. These aramids are used in aerospace and military applications, for ballistic-rated body armor fabric and ballistic composites, in bicycle tires, marine cordage, marine hull reinforcement, and as an asbestos substitute. Kevlar is poly-paraphenylene terephthalamide (see Figure 2.4). It was invented at the DuPont company by Stephanie Kwolek, an American chemist of Polish origin. Kevlar has several unique properties, e.g., it is very strong but very light and flexible. Its strength is due to the alignment of polymer chains. The molecules are held together by hydrogen bonding. Hydrogen bonds are weak but very significant. Kevlar is used for a wide range of applications, e.g., in cryogenic applications and in the fabrication of personal armors, e.g., combat helmets, ballistic face masks, and ballistic vests. It can also be used in the fabrication of parts of smartphones, tires for bikes, goods used for packaging, in shoes, and for the fabrication of wires that can be used for musical instruments [18–23]. A list of its applications is beyond the scope of this book. It can be found on the websites of the companies using it and of the academic institutions that are looking into it for research purposes.

2.4 NANOCOMPOSITES

Up to this point, we have discussed the definition and types of composites. We have seen that composites can be reinforced by particles, also known as particulate composites. It has been observed that the size of the particles significantly affects the overall nature of the composite. For example, Sumita et al. [24] showed that when polyvinylchloride (PVC) was reinforced with ultrafine SiO_2 particles with a diameter in the range of 70–400 Å, Young's moduli, yield, and breaking stress increased with decreasing particle size. The amount of silica particles added to the composite was also significant. The same group of researchers published that when polypropylene (PP) was reinforced with silica (with a diameter in the nanometer and micron range), it was observed that the stiffness constant increased with an increase in filler content when the particles having a diameter in the nanometer scale were used. It was observed that fillers with particle sizes comparable to that of the crystalline region in the polypropylene matrix have a prominent reinforcing effect in the oriented polymer matrix [25]. Similar behavior was observed in Nylon 6 fibers [26]. Several studies showed that composite stiffness, strength, and toughness are strongly affected by particle size, particle/matrix adhesion, and particle loading. This is expected because strength depends on effective stress transfer between the filler and matrix, and toughness/brittleness is controlled by adhesion. Stiffness of composites increases greatly as the size of the particle reduces to nanoscale, which is attributed to the large surface area [27]. These composites are called nanocomposites. Hence, nanocomposites can be described this way:

Nanocomposites contain at least one of the constituents of nanoscale, i.e., less than 100 nm, "or instead the composite structure exhibits nanosized phase separation of the individual components" [28]. Reinforcement of the matrix with nanoparticles improves several properties. Those properties are listed below:

- Mechanical properties, including strength, modulus, and dimensional stability
- Electrical conductivity decreased gas, water, and hydrocarbon permeability
- Flame retardancy
- Thermal stability
- Chemical resistance
- Surface appearance
- Optical clarity

2.5 APPLICATIONS OF NANOCOMPOSITES

We have discussed that nanocomposites exhibit several properties superior to their matrix, hence they can be used in several applications. The applications of nanocomposites are listed below:

- In the fabrication of batteries
- In the fabrication of lightweight sensors
- In the fabrication of windmill blades
- In the fabrication of aircraft components
- In the health sector

Anodes of nanocomposites made of silicon and carbon nanoparticles provide better contact with lithium electrolytes. Hence, lithium-ion batteries have better output when fabricated using anodes of nanocomposites. These days, flexible batteries are also fabricated by using composites of cellulose with nanotubes. Very light and strong materials are fabricated by reinforcing epoxy with carbon nanotube (CNT). These are used in the formation of blades of windmills as large blades can be constructed that can harness more energy.

It was observed that the addition of graphene to epoxy resulted in much better materials than CNT-reinforced epoxy nanocomposites as far as stiffness and strength was concerned. Perhaps bonds between graphene and epoxy is stronger than bonds between CNT and epoxy. These composites are used in the fabrication of aircraft components and windmill blades as well. Efforts are being made to prepare nanocomposites that are fluorescent as well as magnetic. These materials are used during the MRI and surgery of any tumor inside the human body. The magnetic part of the nanocomposite makes the tumor more visible in MRI whereas the fluorescent part makes the tumor more visible during surgery. It was also observed that these composites help in the growth of bone. The complete list of applications of these composites are not described in this chapter. Readers can always find other applications of nanocomposites over various sources to get a better picture [29,30].

2.6 BULK METALLIC GLASS (BMG) NANOCOMPOSITES

We have discussed quite a lot about composites and nanocomposites. Now, back to bulk metallic glasses (BMGs). We humans have prepared BMG nanocomposites of several kinds. Hereafter, we will discuss various aspects of BMG nanocomposites in the following sections.

As discussed in Chapter 1, synthesis of BMG started with the synthesis of a Au/Si amorphous metal alloy using the splat quenching method. This attracted several groups of scientists. They successfully synthesized a range of BMGs, including Fe-, Ni-, and Co-based BMGs; Au-, Ag-, Pd-, and Pt-based BMGs; and Ti-, Cu-, and Zr-based BMGs. These BMGs exhibit several unique properties, e.g., they have low melting temperatures, which allow us to shape them at low temperatures; hence, copper and steel molds can be used continually. Their melts have low viscosities and can easily be filled in molds, and complex structures and shapes can be formed. Apart from these, they have several good mechanical properties as well, e.g., they have good mechanical strength and hardness and hence can provide a good surface finish. These properties make these BMGs ideal candidates for several commercial applications, including springs, optical devices, cell phone cases, biomedical implants, and sporting equipment such as golf clubs. For example, Cu-ZrNi-Al and Zr-Ti-Cu-Ni-Be based BMGs can be used in the fabrication of cell phone cases and golf clubs [31]. As it is said, nothing is perfect in life and neither are these BMGs perfect. BMGs have several limitations (e.g., the fracture toughness and fatigue limit of BMGs are dependent on the content of oxygen present in the BMG). The presence of unwanted phases or partial crystallization affects the overall property of the BMGs. Production of BMGs at an industrial scale involves several challenges, e.g., the possibility of incorporation of oxygen increases when die casted in large crucibles. Incorporation of oxygen and different unwanted phases makes BMGs brittle. Apart from this, BMGs are intrinsically brittle. To overcome all these limitations, several BMG composites (BMGCs) have been synthesized [31]. The basic ideas and fundamental mechanisms of the formation of BMGCs are discussed below.

BMGCs are classified according to basic ideas and the fundamental mechanism of their formation. They are discussed in the following paragraphs under different and relevant subheadings:

1. In-situ composites with in-situ precipitated nanocrystalline phases either via controlled annealing or deformation-induced devitrification (DID), etc. This process is described briefly here:

    One of the strategies to fabricate BMGCs uses the phase-separation property of BMG. A BMG composition having good GFA is chosen. It is then mixed with another material to form the composite. The process can be explained as follows:

    An element "A" based BMG of composition AxByCzDuEv (x + y + z + u + v = 100 at%) can be selected. Now, another element, say "M," with a positive enthalpy of mixing and a miscibility gap is added to this glass batch. A miscibility gap is any part of the mixture—a glass melt, in this case—where the constituent materials are not completely miscible. The part of the glass melt rich in constituent "A" vitrifies to form a glass whereas the part of the glass melt rich in "M" devitrifies, forming crystal. This process leads to the formation of BMGCs in which crystals are embedded into the BMGs. This method is very commonly used for the fabrication of BMGCs. BMGCs of desired properties can be synthesized via this method by using an appropriate element "M" [32].

    Metalloids such as B, C, Si, and P are important components responsible for the formation of Fe-based BMGs. A series of Fe-based BMGCs were synthesized using the concept described above. Fe has a positive enthalpy of mixing with La and Ce, but repulsion exists between Fe, Cu, and La. When Fe was mixed with $La_{32.5}Ce_{32.5}Co_{15}Al_{10}Cu_{10}$, a glass melt was formed but it existed in two phases. Upon cooling, the Fe-rich portion of the glass melt with compositions $CeFe_2$ and Ce $(Fe, Co)_2$ formed glasses. The remaining glass portion with less Fe content solidified to glass afterward. A composite of Fe-rich BMG with Fe-depleted BMG was formed. This technique can be used for the formation of centimeter-sized BMGC samples. However, the presence of metalloids changes soft magnetic properties and plastic deformation at room temperature [32].

    Normally, very thin (~1 mm) aluminium-containing BMGs are fabricated. These can be used only as ribbons. This is a serious limitation. To overcome this, Lu et al. synthesized a series of Al-containing BMGs using the aforementioned technique. Very-high-Al-content (35 at% to 60 at%) glasses, i.e., glasses having Al(Co, Cu)–(La, Ce)–Fe, with 1.2 GPa strength were synthesized. Iron suppresses the precipitation of Al(La, Ce). Samples of 15 mm thickness were achieved [32].

    The strength and ductility of steel is increased by transformation induced plasticity, commonly called TRIP. TRIP is an effect which improves the strength and ductility of steel. This concept was introduced in BMGs as well for the formation of BMGCs to improve the mechanical properties of BMGs. It was observed that the addition of certain elements in conjunction with the desired cooling rate, i.e., the required quenching conditions of the glass melt, facilitates the formation of BMGCs. Most of the austenitic phases used in the formation of BMGCs are metastable at room temperature; for example, B2-CuZr phase is stable above 988 K and tends to decompose into $Cu_{10}Zr_7$ and Zr_2Cu at low temperatures. B2 is the crystalline phase of the material. Further discussion of B2 phase is beyond the scope of this book. Readers can find detailed discussion of crystal structure in various literature; for example, Ref. [33] discusses this aspect of crystalline materials. Now, coming back to BMGCs, the addition of Al in B2-CuZr improves the GFA of the glass melt. It helps the austenite transformation and destabilizes the martensite CuZr phase by decreasing the martensitic transformation temperature. The crystal structure found at high temperatures is the parent phase, often referred to as austenite, and the phase that results from a martensitic transformation is called martensite. The shape memory effect is a direct consequence of a reversible transformation between austenite and martensite. A detailed description of martensitic transformations can be very complex and again beyond the scope of this book. A lot of literature can be found related to this topic. Readers can get a detailed discussion from references [34] and [35]. Now, coming back to BMGCs, we were discussing the effect of the addition of Al in a B2-CuZr system. It was observed that when less than 2% of aluminium was added to the system, the quenching rate had to be adjusted properly for the fabrication of glass as B2-CuZr crystals were still the main competitors. On the other hand, when the amount of aluminium was increased to 3%–8%, Al_2Zr was formed, which hindered the formation of crystalline structures. BMGCs were prepared when the glass-melt was quenched with appropriate cooling rate. Further increase in the amount of aluminium did favor the formation of BMGC as it hindered the formation of glass itself [32].

The morphology and distribution of crystalline phases present in the BMG affects properties of BMGCs significantly. Scientists must keep several parameters under control to synthesize BMGCs of desired properties, e.g., quenching condition and annealing temperature. It is possible to get crystals in the BMG system by controlling the annealing temperature [32].

It was observed that the TRIP-reinforced BMGCs exhibited better mechanical properties. It is believed that the change in properties is due to (i) blocking effect and (ii) martensitic transformation of B2-CuZr. These effects are discussed below in brief.

a. Blocking effect: Blocking effect increases the plasticity of BMGCs. Experimental results suggested that shear bands deflect at the interface between the amorphous phase, which is the matrix of the composite and the crystalline phase, i.e., the reinforcing phase. The crystalline phase becomes the reinforcing phase as it "blocks" the propagation of shear bands, improving the plasticity of the composite. It was also observed that nanometer-sized spherical crystals effectively blocked the propagation of shear bands. Hence, a deeper level of blocking of propagation of shear bands can be expected. This can improve the plasticity and hence the overall mechanical behavior of the composite.

b. Martensitic transformation: It was also observed that TRIP-reinforced BMGC crystals undergo transformation; for example, B2 phase of CuZr transforms into B-19 phase of CuZr. This phenomenon increases the hardness of the reinforcing phase whereas it increases the softness of the amorphous matrix phase. It is believed that the increase in the softness of the glassy matrix is compensated by the hardness of the reinforcing phase [32].

The ductility of BMGCs can be changed by changing the martensitic transformation via stacking fault energy (SFE). SFE has an important role in the mechanical behavior of materials. Experiments demonstrated that minor additions reducing the SFE of B2-CuZr can remarkably improve the tensile ductility and work-hardening capability of the TRIP-BMGCs. Detailed discussion of SFE is beyond the scope of this book. This is discussed at several places in detail. Readers can find discussion on this topic in literature, e.g., Refs. [36–41] can be referred to for this topic [32].

2. Ex-situ composites with BMGs either as reinforcements or as matrix. In several cases, it was found that forming BMGCs via the ex-situ method is easier than with the in-situ method. It is an effective way to increase the ductility of the BMGs. Using the in-situ method, BMGCs of small sizes can be fabricated. Lu et al. fabricated a long wire-shaped BMGC using the ex-situ method. Metallic wire (e.g., tungsten wire) was cleaned, polished, and then dipped in bulk metallic glass melt under a controlled setting. Wires were cooled by passing argon gas. It was possible to feed several wires at a time into the coating machine. It was also possible to weave these coated wires for further structural applications. SEM results suggested the absence of any crystalline phase at the interface of the amorphous matrix and the wire. It was observed that the composite had larger fracture strength and plastic deformation. Hence, by using the appropriate metallic wire with suitable diameter and length, and by coating it with the desired BMG composition, it is possible to tailor the properties of BMGCs [32].

3. BMG composites reinforced by a three-dimensional metallic network: The presence of inhomogeneity in the reinforcing phase reduces its effectiveness in ex-situ BMGCs. It is believed that the rapid propagation of shear bands will be suppressed if the reinforcing material is present homogenously throughout the matrix in every direction. This hypothesis was tested by reinforcing $Ti_{40}Zr_{25}Cu_{12}Ni_3Be_{20}$ with Cu foam. Copper foam with 200–250 μm was taken in a copper mold. The glass melt for the amorphous matrix phase of the aforementioned composition ($Ti_{40}Zr_{25}Cu_{12}Ni_3Be_{20}$ with melting temperature 983°K) was then soaked in the Cu foam (melting point of copper is 1358°K) under a controlled environment. Argon gas was used to create the controlled environment. The mixture was then solidified. The plasticity of the resulting BMGC increased significantly. Perhaps the long-range propagation of shear bands was hindered because of the walls of copper. Copper, being ductile in nature, temporarily blocks the propagation of shear bands but deforms plastically to release the kinetic energy.

Secondary shear bands formed at the interface between the amorphous matrix and the Cu wall because of the elastic mismatch of these materials. This branching of shear bands is perhaps responsible for the improvement of plasticity in the BMGC [32].

4. Multilayered composites with alternating crystalline-amorphous nanolayers can also be fabricated [42].

In the following section, methods of preparation of different composites are discussed.

$(Cu_{50}Zr_{43}A_{17})_{99}Si_1$ (at. %) was melted using an electric arc in a water-cooled copper furnace under a controlled environment. The process of melting with the electric arc is known as arc melting. The arc melting method is frequently used to melt metals. In arc melting, an electric arc is allowed to generate between electrodes either in a controlled environment or a vacuum chamber. The electric arc hits the sample and melts the sample very fast. Zhang et al. used purified argon gas to create a controlled environment for melting the BMG. The suction casting method was used to prepare cylinder- and rod-shaped products. Cylinders of 2 and 3 mm diameter were prepared. Ribbon-shaped samples with a cross-section of 0.02 mm × 1 mm were prepared using the melt casting method. The selected area electron diffraction (SAED) using a transmission electron microscope (TEM) showed that the rod-shaped glass samples of 3 mm diameter contained crystals of 2–10 nm diameter.

It was observed that the BMGC containing crystalline phase has better ductility than the glassy samples without the crystalline phase. The improvement of ductile character was attributed to the presence of nanocrystalline phase in the sample. This clearly shows that the properties of BMGs can be improved by the formation of BMG nanocomposites [43]. Sheng-hui et al. synthesized amorphous $Fe_{75}Mo_5P_{10}C_{8.3}B_{1.7}$ bulk metallic glass using the arc melting method in a titanium metal–gettered argon atmosphere. Casting was done using the suction method. Samples of diameters 1.5 mm and 2 mm were obtained. It was observed that samples of larger diameter, i.e., 2 mm in size, contained crystalline dendrites of α-Fe. The presence of crystals improved the plasticity of the sample. The formation of α-Fe dendrite–reinforced amorphous material was confirmed by SAED studies using a TEM. It is believed that the overall increase in plasticity of the sample can be attributed to the presence of crystal. These nanosized crystals make the amorphous materials a composite of different properties. This opens a door for scientists and industrialists to fabricate BMGCs of desired qualities [44]. Hofmann et al. synthesized a group of BMGs with various compositions. The experiment was conducted with an aim to increase the microstructural toughening and ductility of BMGs. A list of compositions of glasses synthesized are depicted in Table 2.3 [45].

These BMGs were synthesized in two steps. In the first step, the starting materials were taken in elemental form. They were cleansed ultrasonically. They were then melted using arc melting in a titanium-gettered argon environment. In the second step, the ingots thus produced were heated through induction heating in a water-cooled copper boat. The temperature was measured using a pyrometer. In the second step, the sample was treated in a semisolid state, which coarsens the dendrites.

TABLE 2.3
List of Sample Compositions Synthesized by Hoffmann et al.

Composition	Sample IDs as Published
$Zr_{36.6}Ti_{31.4}Nb7Cu_{5.9}Be_{19.1}$	DH1
$Zr_{38.3}Ti_{32.9}Nb_{7.3}Cu_{6.2}Be_{15.3}$	DH2
$Zr_{39.6}Ti_{33.9}Nb_{7.6}Cu_{6.4}Be_{12.5}$	DH3
$Zr_{41.2}Ti_{13.8}Cu_{12.5}Ni_{10}Be_{22.5}$	Vitreloy1
$Zr_{56.2}Ti_{13.8}Nb_{5.0}Cu_{6.9}Ni_{5.6}Be_{12.5}$	LM 2

Source: Hofmann, D.C., et al., *Nature*, 451, 1085–1089, 2008.

Geometry of the sample was dependent upon the geometry of the copper boat. BMGs of 1 cm thickness having a weight of 35 g were obtained. Energy dispersive X-ray spectrometry (EDS) using scanning electron microscope (SEM) analysis suggested that dendrites having body-centered cubic structures were formed. They mainly contained zirconium (Zr), titanium (Ti), and niobium (Nb). SEM analysis also suggested that partition of DH1, DH2, and DH3 by volume fraction yielded 42%, 51%, and 67% dendritic phase in a glass matrix, respectively. These BMGCs were characterized to study several mechanical properties. It was observed that the in-situ BMGC showed better mechanical properties.

The purpose of fabricating these glasses were (i) to increase the toughness of BMG, (ii) to increase the ductile nature of glass, (iii) to incorporate inhomogeneities into the glass matrix so that shear banding can be initiated around the inhomogeneities, and (iv) to avoid crack development in the materials through in-situ BMGC formation [45]. Ling and Courtney first addressed the idea of shear band confinement in metallic glass composites. To achieve this, ribbons of BMG were sandwiched between thin brass plates [46]. Hays et al. synthesized Zr-Ti-Cu-Ni-Be based BMGCs. The reinforcing phase and the matrix phase were synthesized in-situ. The dendrites composed of titanium (Ti), zirconium (Zr), and niobium (Nb) were synthesized in-situ with the bulk metallic glass, which is the matrix phase of the composite. A body-centered cubic (bcc) structure was observed for the dendrites. The presence of dendrites improved the plastic strain to failure, impact resistance, and toughness of the metallic glass [46]. $Zr_{48}Cu_{47.5}Al_4Nb_{0.5}$ was prepared by arc melting using starting materials in elemental form with a purity more than 99.9%. The melt was cast into a copper mold. Samples of 3 mm diameter and 85 mm length were produced. The glass was reinforced with dendrite formed in-situ during the glass formation. The presence of dendrites significantly helps in forming the multiple shear bands, which dramatically improves the plastic deformation capability. These BMGCs also exhibit large elastic strain limit. The presence of crystals was confirmed by SEM images as well as XRD results [47].

Materials with low densities are used for a wide range of applications. Efforts are being made to synthesize several such materials, including aluminum-based

TABLE 2.4

Composition of Al Alloy Used by Eckert et al.

Element	Wt%	Element	Wt%
Al	Base	Fe	0.22
Zn	5.38	Mn	0.24
Mg	2.48	Si	0.15
Cu	1.48	Cr	0.18

Source: Wang, Z., et al., *J. Alloys Compd.*, 651, 170–175, 2015.

(Al-based) matrix composites (AMMCs). They have several unique properties, e.g., low densities and high mechanical strengths. They are used in automotive and aerospace industries and in several defense-related applications. However, AMMCs have several shortcomings, e.g., porosity and interfacial reactions of the materials, which limit their applications. A crack in the material propagates leading fracture or failure. To overcome these defects, Eckert et al. synthesized a composite of Al alloy with BMGs. The composition of the Al alloy is depicted in Table 2.4 [48].

The materials shown in Table 2.4 were gas atomized, with particle size 10 μm taken as starting materials. The metallic glass (MG) with composition $Ti_{52}Cu_{20}Ni_{17}Al_{11}$ was taken for reinforcement. The MG was prepared by the mechanical alloying method using a high-energy planetary ball mill (QM-2SP20, Nanjing NanDa Instrument Plant, Nanjing, China). Al alloy and MG were again mixed and ball-milled in an inert atmosphere. Argon of 99.999% purity was used to achieve an inert environment to minimize any atmospheric contamination. A few drops of ethanol were added to the system to control the process. X-ray diffraction studies showed that the amorphous state of the glassy particles was maintained as only the expected XRD peaks were obtained. The absence of any unexpected peak in the XRD result showed that the material was still glassy after ball-milling the mixture. The shape and the size of the final product was very much dependent upon the number of hours the mixture was ball-milled for the preparation of the composite. According to the literature, in the initial 10 hours, the size of the particles increased with an increase in the milling time. After 10 hours, the composite particles broke, and the sizes of the particles decreased significantly for the next 20 hours. When the samples were further ball-milled for the next 10 hours (i.e., 30 hours of ball-milling), the particle size reduced to ~23 μm. Interestingly, further ball-milling (i.e., more than 30 hours of ball-milling) of the sample does not change the sizes of the particles significantly. The change in particle size with time of ball-milling is shown in [48]. It was concluded that the average microhardness of the composite powders increases with increasing milling time [48].

In another effort, Dembinski et al. synthesized an Al matrix composite with $Cu_{43}Zr_{43}Ag_7Al_7$ in the amorphous state. Amorphous $Cu_{43}Zr_{43}Ag_7Al_7$ was prepared using the gas atomization method. The glass, i.e., the amorphous $Cu_{43}Zr_{43}Ag_7Al_7$, was sieved to collect particles of ~60 μm size. The two powders were mixed in appropriate amounts in a planetary ball mill for 30 minutes in a controlled environment of flowing argon gas. The ball mill was rotated at a speed of 220 rpm (i.e., 220 rotations/minute) for 40 hours. Intermittently, the reaction mixture was cooled down after every 45 minutes after every 15 minutes of milling. It was believed that the cooling exercise avoided overheating of the reaction mixture. SEM analysis showed that the presence of nanocrystalline Al-matrix composites strengthened amorphous powders with different percentages. It was observed that the increase in amorphous phase in the composite increases the products' compression strength as well as hardness, although cracks were observed at the interfaces between the two phases. Partial crystallization of the amorphous phase was observed during hot pressing; however, this phase was equally as strong as the amorphous phase [49].

Much work has been done in this area, and not everything has been reviewed here, although readers can find comprehensive reviews written by several authors. For example, you can refer to the review written by Gupta et al. from Singapore University [50]. Mostly, it was observed that the reinforcement of amorphous materials with other crystalline materials improved the overall mechanical properties of the product, i.e., the composite. The synergistic effect of the matrix material and the reinforcing material can be observed using advanced materials characterization techniques. The enhanced properties of the material make it useful for several applications. In the next section, we will discuss the application of the composites.

A small list of alloys having good glass-forming-ability is listed in Table 2.5.

These BMGs have extremely good mechanical and physical properties, for example, high yield strength, high elastic strain limit in conjunction with high fracture toughness, and fatigue and corrosion resistance. However, these glasses have low tensile ductility. To overcome these limitations, Choi et al. fabricated three sets of alloys by arc melting, as shown in Table 2.6.

The pre-alloyed materials as shown in Table 2.6 were mixed with ceramics like SiC, WC, TiC, and metallic tungsten (W) and tantalum (Ta) as second phase. The mixture was heated via induction heating in a water-cooled copper boat in a Ti-gettered argon atmosphere. Volume fractions in the range of 5%–30% of the second phase were mixed to form the composite. The composite ingots were re-melted in a temperature range of 800°C–1100°C under a vacuum in a quartz tube using induction heating. The melt was then injected through a copper nozzle into an argon atmosphere. It was observed that the addition of reinforcing materials did not affect the glass-forming ability (GFA) of these alloys. The motivations of adding second-phase was to hinder propagation of shear bands [51].

TABLE 2.5
List of Alloys Having Good Glass-Forming Ability (GFA)

S. No.	Name of Metals Present in Glass-Forming Alloys as Published [ref.]
1	Lanthanum, Aluminium, Nickel
2	Zirconium, Aluminium, Copper, Nickel
3	Zirconium, Titanium, Copper, Nickel, Beryllium
4	Zirconium, Titanium, Copper, Nickel
5	Zirconium, Titanium (Niobium), Aluminium, Copper, Nickel

TABLE 2.6

Compositions of Alloys Fabricated by Choi et al.

S. No.	Composition	Method of Alloy Formation	Purity Level
1	$Cu_{47}Ti_{34}Zr_{11}Ni_8$	Arc melting	>99.7%
2	$Zr_{52.5}Ti_5Al_{10}Cu_{17.9}Ni_{14.6}$	Arc melting	>99.7%
3	$Zr_{57}Nb_5Al_{10}Cu_{15.4}Ni_{12.6}$	Arc melting	>99.7%

Source: Choi-Yim, H., and Johnson, W.L., *Appl. Phys. Lett.*, 71, 3808–3810, 1997.

Several groups of scientists fabricated BMGCs to improve the mechanical properties of BMGs, such as tendency of deformation twinning (a phenomenon where crystals deform plastically). In deformation twinning, each row of the crystal is displaced slightly by sliding along the planes. To overcome these shortcomings, Lu et al. reinforced BMGs with a range of metals like Co, Ti, Fe, Ni, Ta, Cr, Ga, Hf, Nb, Ta, and Ag. Alloys with a nominal composition of $Zr_{48}Cu_{48-x}Al_4M_x$, where $x = 0$ to 2 at% and M is one of the aforementioned metals, were prepared via the arc melting method. Starting materials of 99.9% purity were used. The alloy ingots were prepared in a Ti-gettered argon atmosphere. The samples were melted six times for homogenization. Homogenized samples were molded using Cu molds. High-performance BMGCs were formed with this method [52].

Throughout the chapter, the different applications of BMGCs are discussed. They are and can be used in wide applications. They can be used in the fabrication of various parts required for the construction of automobiles and airplanes and in defense. They can be extensively used in biomedical applications and 3D printing as well.

2.7 SUMMARY

As the name says, composites are materials composed of two different materials. The difference between a solution and a composite is that, in a solution, we can differentiate the existence of the two materials whereas, in composites, the constituents exist in two different phases, retaining their identities. They exist together and work together but retain their identities! In a composite, one or more reinforcing materials is present in a base material called matrix material. Concrete is a composite in which cement is the matrix material and stone chips are mixed in as the reinforcing material. The matrix material, which is cement in this case, unloads the weight on the stone chip. All high-rise buildings are constructed using concrete. The term "concrete jungle" is coined from there.

Several types of composites are synthesized based on the rising demand. For example, we have natural fiber–reinforced composites, glass fiber–reinforced composites, composites reinforced with a variety of particles, etc. These are discussed in brief in this chapter. These composites are used for several important applications, e.g., in the fabrication of several parts of automobiles and airplanes, for biomedical applications, etc. Bulk metallic glasses are attractive materials for several specific applications. The application of these glasses is limited because of a few undesirable properties, e.g., brittleness. To overcome this, glasses are reinforced with nanocrystalline phase materials. There are several methods for preparing these composites. These are categorized as in-situ and ex-situ. It was always observed that the properties of the resulting composite were better than the base bulk metallic glasses.

REFERENCES

1. https://www.alamy.com/stock-photo-indian-women-hard-working-at-construction-site-shoveling-gravel-making-14785037.html, Last accessed 27 February 2019.
2. https://encrypted-tbn0.gstatic.com/images?q=tbn:ANd9GcRDCS3UyZoTcBJ7-QEzmmfCg5nj6eGzAqF3ovvOFb4AVeRRZkCv, Last accessed 27 February 2019.
3. https://www.concretenetwork.com/concrete-prices.html, Last accessed 27 February 2019.
4. https://www.instructables.com/id/Giant-Concrete-Buddha-Head-Garden-Sculpture/, Last accessed 27 February 2019.
5. https://en.wikipedia.org/wiki/Saint_Sophia%27s_Cathedral,_Kiev, Last accessed 27 February 2019.
6. https://www.dreamstime.com/stock-photo-pont-du-gard-france-image7428200, Last accessed 27 February 2019.
7. SEM image of cork taken by the author (Dr. Nupur Prasad), Brunel University, London, UK.
8. Boyde, A. (2011), "Scanning electron microscopy of bone," *Bone Research Protocols*, pp. 365–400. doi:10.1007/978-1-61779-415-5_24.
9. https://nptel.ac.in/courses/Webcourse-contents/IIScBANG/Composite%20Materials/pdf/Teacher_Slides/mod1.pdf, Last accessed 17 February 2019.
10. https://www.reddit.com/r/pics/comments/4563qh/pearls_in_an_oyster/, Last accessed 28 February 2019.
11. Rawal, S. (2001), "Metal-matrix composites for space applications," *Journal of Materials*, Vol. 53, No. 4, pp. 14–17.
12. Koniuszewska, A.G. and Kaczmar, J.W. (2016), "Application of polymer based composite materials in transportation, progress in rubber," *Plastics and Recycling Technology*, Vol. 32, No. 1, pp. 1–24.
13. Park, S.J. and Seo, M.K. (2011), "Types of composites." *Interface Science and Technology*, Vol. 18, pp. 501–629.
14. https://www.racheljetel.com/recent-projects/2017/7/27/hong-kong-exploring-a-more-sustainable-concrete-jungle, Last accessed 24 February 2019.
15. https://www.google.com/search?q=images+of+concrete+jungle&tbm= isch&source= univ&sa=X&ved=2ahUKEwjC3OvltPgAhXSZSsKHQKgA9wQsAR6BAgAEAE& biw=1366&bih=625#imgrc=LQmllgHs1YJd8M: Last accessed 24 February 2019.
16. https://www.olx.co.ke/item/boat-diving-boat-made-of-fiberglass-price-for-810-people-iid-1051462113; Last accessed 24 February 2019.
17. https://www.shutterstock.com/image-photo/fiberglass-surface-tissue-thin-fiber-glass-1207735696?src=0dH3uAEQE36Ng2tGd24Kkg-2-9, Last accessed 24 February 2019.
18. https://www.google.com/search?q=Airbus+A350+decorated+in+carbon+fibre&tbm= isch&source=univ&sa=X&ved=2ahUKEwjEgd-TvuPgAhVXWysKHcreCPYQsA R6BAgAEAE&biw=1366&bih=625#imgrc=YjsrI58GGjTNKM, Last accessed 24 February 2019.
19. http://tech-racingcars.wikidot.com/citroen-sm-serie-sb, Last accessed 24 February 2019.
20. Reashad, B.K. and Nasrin, F. (2012), "Kevlar-the super tough fibre," *International Journal of Textile Science*, Vol. 1, No. 6, pp. 78–83.
21. https://en.wikipedia.org/wiki/Kevlar, Last accessed 24 February 2019.

22. http://www.essentialchemicalindustry.org/materials-and-applications/composites. html, Last accessed 24 February 2019.
23. https://kevlarweb.wordpress.com/applications, Last accessed 24 February 2019.
24. Sumita, M., Ookuma, T., Miyasaka, K., Ishikawa, K. (1984), "Mechanical properties of oriented polyvinylchloride composites filled with ultrafine particles," *Colloid & Polymer Science*, Vol. 262, pp. 103–109.
25. Sumita, M., Ookuma, T., Miyasaka, K., Ishikawa, K. (1982), "Effect of ultra-fine particles on the elastic properties of oriented polypropylene composites," *Journal of Materials Science*, Vol. 17, pp. 2869–2877.
26. Sumita, M., Shizuma, T., Miyasaka, K., and Ishikawa, K. (1983), "Effect of reducible properties of temperature, rate of strain, and filler content on the tensile yield stress of nylon 6 composites filled with ultrafine particles," *Journal of Macromolecular Science and Physical Science*, Vol. 22, No. 4, pp. 601–618.
27. Fu, S.Y., Feng, X.Q., Lauke, B., Mai, Y.W. (2008), "Effects of particle size, particle/matrix interface adhesion and particle loading on mechanical properties of particulate–polymer composites," *Composites: Part B*, Vol. 39, pp. 93–96.
28. https://www.msm.cam.ac.uk/research/research-disciplines/composite-and-nanocomposite-materials, Last accessed 24 February 2019.
29. http://www.understandingnano.com/self-assembled-nanocomposites-battery-anode. html, Last accessed 24 February 2019.
30. https://www.shutterstock.com/search/windmill+blades, Last accessed 24 February 2019.
31. Hofmann, D. (2013), "Review article bulk metallic glasses and their composites: a brief history of diverging fields," *Journal of Materials*, Vol. 2013, p. 517904.
32. Wu, Y., Wang, H., Liu, X.J., Chen, X.H., Hui, X.D., Zhang, Y., Lu, Z. (2014), "Designing bulk metallic glass composites with enhanced formability and plasticity," *Journal of Materials Science & Technology*, Vol. 30, No. 6, pp. 566–575.
33. https://nptel.ac.in/courses/113108052/module4/lecture23.pdf, Last accessed 6 April 2019.
34. Olson, G.B. and Owen, W.S. (1992), *Martensite*, ASM International, Materials Park, OH.
35. https://nitinol.com/reference/the-thermal-transformation-from-austenite-to-martensite-and-the-origin-of-shape-memory/, Last accessed 4 April 2019.
36. https://nptel.ac.in/courses/113105023/Lecture13.pdf, Last accessed 6 April 2019.
37. https://nptel.ac.in/courses/113102080/61, Last accessed 6 April 2019.
38. Hirth, J.P. (1970), "Thermodynamics of stacking fault," *Metallurgical Transactions*, Vol. 1, pp. 2367–2374.
39. Hirth, J.P. and Lothe, J. (1968), *Theory of Dislocations*, McGraw-Hill Book, New York.
40. Darken, L.S. and Gurry, R.W. (1953), *Physical Chemistry of Metals*, McGraw-Hill Book, New York.
41. Fast, J.D. (1962), *Entropy*, McGraw-Hill Book, New York.
42. Khan, M.M., Nemati, A., Rahman, Z.U., Shah, U.H., Asgar, H., Haider, W. (2017), "Recent advancements in bulk metallic glasses and their applications: A review," *Critical Reviews in Solid State and Materials Science*, Vol. 43, No. 3, pp. 233–268.
43. Malekana, M., Shabestari, S.G., Zhang, W., Seyedein, S.H., Gholamipourd, R., Yubuta, K., Makinoc, A., Inoue, A. (2012), "Formation of bulk metallic glass in situ nanocomposite in $(Cu_{50}Zr_{43}Al_{7})_{99}Si_1$ alloy," *Materials Science and Engineering A*, Vol. 553, pp. 10–13.
44. Sheng-feng, G., Jing-feng, W., Hong-ju, Z., Sheng-hui, X. (2012), "Enhanced plasticity of Fe-based bulk metallic glass by tailoring microstructure," *Transition Nonferrous Materials Society China*, Vol. 22, pp. 348–353.
45. Hofmann, D.C., Suh, J., Wiest, A., Duan, G., Lind, M., Demetriou, M.D., Johnson, W.L. (2008), "Designing metallic glass matrix composites with high toughness and tensile ductility," *Nature*, Vol. 451, pp. 1085–1089.

46. Hays, C.C., Kim, C.P., Johnson, W.L. (2000), "Microstructure controlled shear band pattern formation and enhanced plasticity of bulk metallic glasses containing in-situ formed ductile phase dendrite dispersions," *Physical Review Letters*, Vol. 84, No. 13, pp. 2901–2904.

47. Wu, F., Chan, K.C., Jiang, S., Chen, S., Wang, G. (2014), "Bulk metallic glass composite with good tensile ductility, high strength and large elastic strain limit," *Scientific Reports*, Vol. 4, No. 1, p. 5302.

48. Wang, Z., Scudino, S., Stoica, M., Zhang, W., Eckert, J. (2015), "Al-based matrix composites reinforced with short Fe-based metallic glassy fibre," *Journal of Alloys and Compounds*, Vol. 651, pp. 170–175.

49. Dutkiewicz, J., Rogal, Ł., Wajda, W., Kukuła-Kurzyniec, A., Coddet, C., Dembinski, L. (2015), "Aluminum matrix composites strengthened with CuZrAgAl amorphous atomized powder particles," *Journal of Materials Engineering and Performance*, Vol. 24, No. 6, pp. 2266–2273.

50. Subramanian, J., Seetharaman, S., Gupta, M. (2015), "Processing and properties of aluminum and magnesium-based composites containing amorphous reinforcement: A review," *Metals*, Vol. 5, No. 2, pp. 743–762.

51. Choi-Yim, H. and Johnson, W.L. (1997), "Bulk metallic glass matrix composites," *Applied Physics Letters*, Vol. 71, No. 26, pp. 3808–3810.

52. Wu, Y., Zhou, D.Q., Song, W.L., Wang, H., Zhang, Z.Y., Ma, D., Wang, X.L., Lu, Z.P. (2012), "Ductilizing bulk metallic glass composite by tailoring stacking fault energy," *Physical Review Letters*, Vol. 109, No. 24, pp. 245506-1–5.

3 Molecular Modeling of Metallic Glasses and Their Nanocomposites

Sumit Sharma, Pramod Kumar, Rakesh Chandra, and Nitin Thakur

CONTENTS

3.1 INTRODUCTION

In the last two decades [1], significant research has been devoted to carbon nanotubes (CNTs) and graphene-reinforced nanomaterials. Graphene, a one-atom-thick sheet of carbon atoms, has aroused considerable interest around the world because of its extraordinary mechanical, electronic, and thermal properties [2]. Graphene sheets, because of their excellent Young's modulus and mechanical strength [2], are good candidates as nanoreinforcement for improving the mechanical properties of polymer-based composites. In contrast with CNT or buckyball, graphene atoms are accessible from both sides, so a stronger interaction with surrounding molecules is created. Experimental studies [3,4] have shown that the single-layer graphene (SLG) can present exceptional thermal conductivity of about 4100 ± 500 W $(mK)^{-1}$, which outperforms all other known materials. This observation proposes the graphene for micro/nanoscale electronic device heat transfer blocks in response to thermal management concerns in the electronics industry.

CNTs are excellent reinforcement because of their unique mechanical properties and large surface area per unit volume. Experiments and calculations show that

CNTs have a modulus equal to or greater than the best graphite fibers, and strengths at least an order of magnitude higher than typical graphite fibers. For example, Yu et al. [5] measured the tensile properties of individual multiwalled carbon nanotubes (MWCNTs) and obtained values of 11–63 GPa for the tensile strength and 270–950 GPa for Young's modulus. For comparison, the modulus and strength of graphite fibers are 300–800 and 5 GPa, respectively. In addition to their outstanding mechanical properties, the surface area per unit volume of CNTs is much larger than that of embedded graphite fibers. For example, 30 nm diameter nanotubes have 150 times more surface area than 5 μm diameter fibers for the same filler volume fraction, such that the nanotube/matrix interfacial area is much larger than that in traditional fiber-reinforced composites. The unusual mechanical strength of the CNTs has motivated scientists to fabricate and modify other useful materials which are cheaply available in bulk form by combining them as composites with CNTs. For several years, graphene and CNTs have been used as reinforcements for polymeric composites because of their relative ease of processing. However, in recent times, there has been increased interest in using matrices such as metals, ceramics, and semiconductors in addition to that of polymers. This study deals with metallic glass (MG) nanocomposites.

MG is defined as a material which can retain the disordered atomic structure of high-temperature melts. Figure 3.1 shows the amorphous structure of $Cu_{64}Zr_{36}$ alloy having 5233 Cu atoms and 2956 Zr atoms. MG composites do not contain crystallographic defects, such as grain boundaries and dislocations. In the early 1970s, metallic glasses, such as amorphous ribbons and wires, mainly included magnetic Fe- and Co-based ribbons, which were focused on their excellent softer magnetic properties. Chen et al. [6] investigated the magnetic properties of zero magnetostrictive metallic glass using both standard flux-meter methods and domain patterns obtained

Copper (Cu)   Zirconium (Zr)

FIGURE 3.1 Amorphous structure of $Cu_{64}Zr_{36}$ alloy having a Zr and Cu molar ratio of 1:1.7.

from a scanning electron microscope technique. It was shown that the material was not magnetically isotropic but rather had preferred directions of magnetization that were determined by the details of the quenching process.

In the 1980s, more and more MGs were developed in various alloy systems, and there were considerable efforts on the fundamental understanding of structural, atomic, and electronic transport properties and low temperature behavior as well as further exploration of mechanical, magnetic, and chemical properties [7]. Cylindrical samples with diameters of up to 5 mm or sheets with similar thicknesses were made fully glassy by casting $La_{55}Al_{25}Ni_{20}$ [8]. In the early 1990s, the $Zr_{65}A_{17.5}Ni_{10}Cu_{17.5}$ MGs were fabricated by water quenching with a diameter up to 16 mm [9]. In 2012, the world's largest glassy alloy ever made was $Pd_{42.5}Cu_{30}Ni_{7.5}P_{20}$, which can be cast into amorphous rods of 80 mm diameter by copper mold casting [10].

In the twenty-first century, MGs have been researched with significant enthusiasm, driven by both a fundamental interest in the structures and properties of disordered materials and their unique promise for structural and functional applications. MGs are the potential candidates in the structural engineering field owing to their impressive suite of mechanical properties [11]. MGs exhibit high strengths and even an ultrahigh strength of over 5 GPa [11], high hardness, high specific strengths, superior elastic limits (2%), and high scratch and wear resistances. However, the lack of the macroscopic plasticity or ductility severely limits their structural applications at room temperature. MGs have a large region of elastic behavior but absorb a low energy when subjected to a strain loading. Many efforts have been made in order to increase their energy storage ability and therefore improve their ductility characteristics.

Kawamura et al. [12] investigated the deformation behavior of $La_{55}Al_{25}Ni_{20}$ MG that had a wide supercooled liquid region of 72 K before crystallization. The supercooled liquid exhibited high strain rate superplasticity. The MG showed large elongations of more than 1000% at strain rates ranging from 10^{-4} to $10^{0}\,s^{-1}$ and at low temperatures of about 0.7 T_m and retained the ductile nature without crystallization even after deformation. Wang et al. [13] prepared the CNT-reinforced, Zr-based bulk MG composites using the conventional die cast method. Investigation showed that the elastic moduli, Vicker's hardness, density, and Debye temperature of the composite were markedly changed by introducing CNTs into the glass. All the parameters increased with the introduction of CNTs in MG except density, which showed a decreasing trend. Mukai et al. [14] analyzed the tensile behavior of a bulk MG $Pd_{40}Ni_{40}P_{20}$ under both quasi-static and dynamic strain rate conditions. No major difference was observed. Multiple shear bands were formed in samples tested at the dynamic strain rate. However, shear band interaction appeared to have an insignificant effect on the plasticity of the alloy. Tensile fracture stress of the material was found to be very high (about 1600 MPa) and was virtually independent of strain rate.

Ravichandran et al. [15] studied the stress-strain relations for the $Zr_{41.2}Ti_{13.8}$ $Cu_{12.5}Ni_{10}Be_{22.5}$ bulk MG (Vitreloy 1) over a broad range of temperatures (room temperature to its supercooled liquid region) and strain rates (10^{-5} to $10^{3}\,s^{-1}$) in uniaxial compression using both quasi-static and dynamic Kolsky (split Hopkinson) pressure bar loading systems. Relaxation and jump in strain rate experiments were conducted to further understand the time-dependent behavior of Vitreloy 1. The material exhibited superplastic flow above its glass transition temperature (623 K) and strain rates of up to 1 s^{-1}. The viscosity in the homogeneous deformation regime was found to decrease dramatically with increasing strain rate. Johnson and Samwer [16] obtained the room temperature elastic constants and compressive yield strengths of 30 metallic glasses. The average shear limit was found as: $\gamma_C = 0.0267 \pm 0.0020$, where $\tau_Y = \gamma_C G$ was the maximum resolved shear stress at yielding, and G the shear modulus. Li and Li [17] investigated the mechanical response of amorphous metal $Ni_{40}Zr_{60}$ under applied tensile loading using large-scale atomistic simulations. It was found that the notch-free samples showed strength close to the theoretical fracture strength and extremely large ductility. Apparent strain hardening and strain rate sensitivity were also observed in these studies. It was argued that the free volume generation and localized shear displacement were responsible for the mechanical properties.

Sui et al. [18] used in-situ tensile tests in a transmission electron microscope to demonstrate radically different deformation behavior for monolithic metallic-glass samples with dimensions of the order of 100 nm. Large tensile ductility in the range of 23%–45% was observed, including significant uniform elongation and extensive necking or stable growth of the shear offset. Ramamurty et al. [19] conducted uniaxial compression experiments on 0.3, 1, and 3 μm diameter micropillars of a Zr-based bulk metallic glass in as-cast, shotpeened, and structurally relaxed conditions. Shear band formation and stable propagation was observed to be the plastic deformation mode in all cases, with no detectable difference in yield strength according to either size or condition. Jang and Greer [20] showed that when reduced to 100 nm, Zr-based metallic glass nanopillars attain ceramic-like strengths (2.25 GPa) and metal-like ductility (25%) simultaneously.

Hosson et al. [21] fabricated and tested nanosized pillars with diameters ranging from 90 to 600 nm of four amorphous alloys, $Cu_{47}Ti_{33}Zr_{11}Ni_6Sn_2Si_1$, $Zr_{50}Ti_{16.5}Cu_{15}Ni_{18.5}$, $Zr_{61.8}Cu_{18}Ni_{10.2}Al_{10}$, and $Al_{86}Ni_9Y_5$, in a transmission electron microscope. The yield stress of all the MGs measured through the in-situ experiments was found to be essentially size independent, irrespective of tapering. With increasing size, all the MGs examined showed a ductile-to-brittle transition under compression; the transition point, however, depended on the chemical composition of the specific MG investigated. Misra et al. [22] fabricated CNT-reinforced $Mg_{59.5}Cu_{22.9}Gd_{11}Ag_{6.6}$ bulk MG composites using the approach of differential pressure casting in a copper mould. The presence of CNT in the amorphous matrix enhanced the compressive strength and fracture strain, and the compressive strength and fracture strain of a composite with 3 vol.% CNT approached approximately 1007 MPa and 0.42%, respectively. In composites with less than 3 vol.% CNT, the glass-forming ability was similar to the un-doped MG. Gupta et al. [23] synthesized and characterized $Ni_{60}Nb_{40}$ amorphous alloy particle-reinforced Mg composites with varying volume fractions. Microwave-assisted two-directional rapid sintering technique followed by hot extrusion was used to produce these pure Mg-based composites. The composites showed significant improvement in hardness (increment up to 120%) and compressive strength (~85% increase at 5 vol.%).

Kalcher et al. [24] performed molecular dynamics (MD) simulations on the creep behavior of $Cu_{64}Zr_{36}$ metallic glass composites. The results revealed that all composites exhibited much higher creep rates than the homogeneous glass. This was because the glass-crystal interface acted like a weak interphase, where the activation of shear transformation zones was easier than in the surrounding glass. We observed that the creep behavior of the composites did not only depend on the interface area but also on the orientation of the interface with respect to the loading axis. The authors proposed an explanation in terms of different mean Schmid factors of the interfaces, with the amorphous interface regions acting as preferential slip sites.

Controlling and manipulating heterogeneity through spinodal decomposition is one of few effective ways in making better metallic glass matrix composites. Wang et al. [25] used a finite element modeling approach to investigate how the geometry, orientation, size, and statistical and spatial distribution of varying free volume heterogeneities in spinodal decomposition affected the mechanical properties of the composites. Among the plethora of factors, orientation and statistical distribution of free volumes in the spinodal microstructures were identified as two key ones critically influencing the mechanical responses. The size effect was also discovered and found to be governed by the minimum size in shear banding initiation and propagation. Different amounts of nitrogen from 1000 to 14,000 wt. ppm were added into the $Ti_{48}Zr_{20}Nb_{12}Cu_5Be_{15}$ bulk metallic glass composites by Li et al. [26]. Results showed that the addition of nitrogen can increase the yield strength of composites from 1400 to 2350 MPa due to the solid solution strengthening effect. The composites still owned a certain plastic strain (above 10% when they contain 14,000 ppm nitrogen) before fracture under compressive tests. The reason was attributed to nitrogen being a much weaker activator for the formation of brittle α-Ti phase compared with oxygen, which made the β-Ti dendrite still softer than that of the glassy phase, thus enhancing the strength of the materials, but it did not deteriorate the plasticity too much. The research showed that nitrogen could be added into the Ti-based metallic glass composites in a very large range to effectively tune the suitable mechanical properties of Ti-based metallic glass composites, without worrying about the catastrophic failure in the process of deformation.

Lee et al. [27] reported the preparation and wear behavior of mechanically alloyed Ti-based bulk metallic glass composites containing carbon particles. The carbon/$Ti_{50}Cu_{28}Ni_{15}Sn_7$ metallic glass composite powders were formed by a two-stage mechanical alloying (MA) process. The bulk metallic glass composite was successfully prepared by vacuum hot pressing of as-milled carbon/$Ti_{50}Cu_{28}Ni_{15}Sn_7$ metallic glass composite powders. The differential scanning calorimeter (DSC) results showed that the thermal stability of the amorphous matrix was affected by the presence of the carbon particles. Changes in Tg and Tx suggested deviations in the chemical composition of the amorphous matrix that were developed due to a partial dissolution of the carbon species into the amorphous phase. Although the hardness of carbon/$Ti_{50}Cu_{28}Ni_{15}Sn_7$ bulk metallic glass composite was increased with carbon addition, their wear resistance was not directly proportional to hardness and did not followed the standard wear law. The increase in the wear rate of a carbon/$Ti_{50}Cu_{28}Ni_{15}Sn_7$ bulk metallic glass composite with high carbon content might be attributed to the residual porosity and the hard TiC particles formed in-situ in the matrix of the composite.

The antimicrobial and wear behavior of metallic glass composites correspond-ing to the $Cu_{50+x}(Zr_{44}Al_6)_{50-x}$ system with $x = (0, 3,$ and $6)$ was studied by Villapun et al. [28]. The three compositions consisted of crystalline phases embedded in an amorphous matrix and exhibited increased crystallinity with increasing Cu content, i.e., a decrease of the glass-forming ability. The wear resistance also increased with the addition of Cu as indirectly assessed from H/Er and H^3/Er^2 parameters obtained from nanoindentation tests. These results were in agreement with scratch tests since the alloy with the highest Cu content, i.e., $Cu_{56}Zr_{38.7}Al_{15.3}$, revealed a crack increase, lower pile-up, prone adhesion wear in dry sliding, and higher scratch groove volume to pile-up volume. Samples with the higher Cu content revealed higher hydrophilicity. Time-kill studies revealed higher reduction in colony-forming units for *Escherichia coli* (gram-negative) and *Bacillus subtilis* (gram-positive) after 60 minutes of contact time for the $Cu_{56}Zr_{38.7}Al_{15.3}$ alloy, and all the samples achieved a complete elimination of bacteria in 250 minutes.

A novel kind of Zr-based bulk metallic glass composites (BMGCs) reinforced with tailored volume fractions of tungsten springs were designed by Chen et al. [29]. The effect of tungsten springs on the compressive mechanical properties and fail-ure modes of the BMGCs was investigated. When the volume fraction of tungsten springs was increased, the yield strength decreased slightly; however, both the frac-ture strength and plastic strain increased significantly. The continuity and special geometric shape of the tungsten springs were revealed to account for the enhanced fracture strength and plasticity. The BMGCs with lower volume fractions of tungsten springs exhibited a unique cone-shaped fracture morphology due to the preferential initiation of shear bands in the central metallic glass matrix while the BMGCs with larger volume fractions of tungsten springs exhibited a mixed shear fracture mor-phology consisting of four distinct regions. Nevertheless, all the current BMGCs failed in a shear mode, which was obviously different from that of the BMGCs rein-forced with longitudinal fibers. The failure modes of the BMGCs could be modified by designing the structure and orientation of the tungsten springs.

$Ti_{45}Cu_{40}Ni_7Zr_5Sn_{2.5}Si_{0.5}$ alloys were prepared by Hong et al. [30] under various cooling rate conditions during solidification. The alloys exhibited different volume fractions of B2 particles with $0 \sim 40$ vol.% in an amorphous matrix. Monolithic bulk metallic glass of 1 mm diameter showed no macroscopic plasticity and exhib-ited the typical vein patterns in a maximum shear stress plane on the fracture sur-face. However, a bulk metallic glass composite containing the B2 particles revealed obvious plasticity ($\sim$5.6%) with a remarkable work-hardening behavior that resulted from a stress-induced martensitic transformation of the B2 particles. Moreover, the composite displayed the complicated fracture morphologies containing three types of fracture features. Through detailed investigations of the microstructural evo-lution, mechanical, deformation, and fracture characteristics, the influence of B2 particles on the overall behavior of the TiCu-based bulk metallic glass composites was elucidated.

The existence of second phase usually promotes the formation of multiple shear bands under uniaxial loading, leading to the apparent strain hardening and enhanced plasticity in bulk metallic glasses (BMGs). However, the effect of second phase on the fracture behavior of BMGs has been rarely investigated when shear banding is suppressed. Pan et al. [31] investigated the effect of Ta particles on the fracture behavior of notched bulk metallic glass composites (BMGCs), where shear banding was suppressed due to the introduction of a high triaxial stress state. With an increase in the volume fraction of Ta particles from 0% to 3.3%, the fracture strength significantly decreased from 2.70 to 1.20 GPa. Different from the veinlike patterns caused by shear banding, the fracture surface of notched BMGCs displayed numerous equiaxed dimples and micro-cracks, indicating that voids nucleation and coalescence govern the fracture process. Due to the relatively weak bonding of the interface and the mechanical incompatibility between Ta particles and amorphous matrix, voids nucleated at the interface during tensile deformation, giving rise to a reduction in the fracture strength of BMGCs.

One possible key to improve plasticity in a CuZrAl bulk metallic glass is to introduce soft metallic crystalline particles. Cardinal et al. [32] used tantalum particles with two different geometries: (i) irregularly shaped particles of equivalent average diameter of 45 μm (Ta1) and (ii) spherical particles with an average size of 25 microns (Ta2). When the volume fraction of tantalum was increased from 0% to 50%, the modulus increased (from 95 to 140 GPa), the Poisson's ratio decreased only slightly (from 0.37 to 0.35), and the hardness decreased regularly (from 485 HV30 to about 300 HV30), whatever the particle size. All these results were in agreement with simple laws of mixtures, involving only the volume fraction of the second phase. The results obtained during the compression tests were more complicated. The yield strength decreased when the Ta content exceeded 20% in volume. This evolution was identical for both types of particles. Similarly, an increase in plasticity was always observed when the tantalum content exceeded about 10% in volume. For the highest values of this content, a significant damage of the sample followed this plastic deformation before rupture. But a large dispersion of the values was observed, more particularly for small particles of tantalum. This was due to a problem of heterogeneity of the particle dispersion and the presence of clusters; this difficulty in dispersing the particles was more marked for small spherical particles. An image analysis was performed to characterize this clustering effect. Clusters had a negative effect because they facilitated the propagation of cracks through Ta-Ta weak interface instead of deflecting it. An optimal content to combine a high elastic limit, a good hardness, and a high plasticity seemed to exist near 30% vol. of crystalline particles.

The above studies suggest the promising reinforcement role of CNTs in metallic glasses. These studies motivated the authors to investigate the role of nanofiller geometry in MG composites. While experimental research shows increment in mechanical properties of MG by adding CNTs, it is important to explore the basic mechanisms of CNT-MG interaction during mechanical loading. It is also of particular importance to investigate how CNTs and SLGS influence the yielding mechanisms of MGs at atomic scale, which motivated this research by using molecular dynamics (MD) simulations. This research will contribute to the design of MG-based nanocomposites with improved strength and ductility simultaneously.

To the best of the knowledge of the authors, this will be the first study to examine the role of CNTs and SLGs in MG matrix. This study utilized MD simulations to predict the Young's moduli of MG nanocomposites reinforced with unidirectional CNT and graphene. The remaining study is organized as follows. Section 3.2 outlines the computational model and the details of MD simulation. MD results are presented and discussed in Section 3.3. Finally, Section 3.4 concludes with a summary of the results obtained.

3.2 MD SIMULATION

3.2.1 Inter-Atomic Potential

MD simulations were performed to obtain the mechanical properties of MG reinforced with CNTs and graphene. Materials Studio 7.0 molecular modeling software from Biovia was used to perform the simulations. With MG as the matrix material, four case studies involving pure SLGS, pristine MG, a SLGS/MG sample, and CNT/MG nanocomposite were investigated under uniaxial tensile loadings. First, the graphene sample was analyzed and its Young's modulus was compared with the corresponding values available in the literature. In the next stage, the pristine MG sample was studied. Finally, the significance of the nanofiller geometry in reinforcing MG composites was explored by analyzing two additional representative volume elements (RVEs) under identical conditions. As discussed in the next section, to have a better comparison, these RVEs include two distinct nanofillers, i.e., SLGS and CNT, with the same V_f.

MD simulations were performed using the condensed phase optimized molecular potential for atomistic simulation studies (COMPASS) force field. COMPASS is a general all-atom force field for atomistic simulation of common organic molecules, inorganic small molecules, and polymers, developed by using state-of-the-art ab initio and empirical parameterization techniques. It has been validated for a wide variety of systems, including metals. This force field has been widely used to study the properties of CNTs, polymers, and CNT/polymer composites by means of MD calculations [33,34].

3.2.2 MD Simulation Setup

Firstly, an amorphous configuration of $Cu_{64}Zr_{36}$ was created using the "amorphous cell" module of Materials Studio 7.0. Figure 3.1 shows the structure of a $Cu_{64}Zr_{36}$ system having dimensions of 50 Å × 50 Å × 50 Å and containing a total of 8190 atoms. Periodic boundary conditions were applied in all the three directions. In order to create an equilibrated structure, the MG sample was first heated to a temperature of about 2000 K. The number of annealing cycles was taken as 5. It was then equilibrated for 100 ps. The MG specimen was then quenched to 10 K. The time step was taken as 1 fs and the total simulation time for the dynamics was 100 ps. Figure 3.2 shows the radial distribution function (RDF) of the quenched $Cu_{64}Zr_{36}$ sample, showing the amorphous structure of the MG system. Figure 3.3 displays the

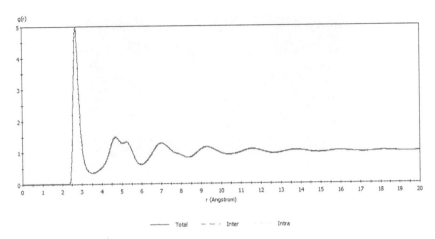

FIGURE 3.2   Radial distribution function (RDF) of the $Cu_{64}Zr_{36}$ specimen after quenching.

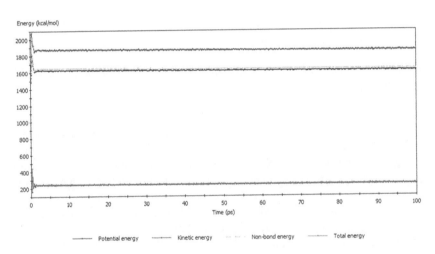

FIGURE 3.3   Energies obtained for the $Cu_{64}Zr_{36}$ specimen after dynamics run during quenching.

energies obtained for the $Cu_{64}Zr_{36}$ specimen after dynamics run during quenching. The energy after quenching of the MG specimen was found to be 1368 kcal/mol.

In the second step, the graphene sheet and (10,10) armchair CNT were constructed using the "build nanostructure" tool of Materials Studio 7.0. Figure 3.4 highlights the initial configuration of the graphene sample having dimensions of 50 Å × 50 Å and also of (10,10) armchair CNT having a length of 49.18 Å. These nanostructures were then packed inside the MG matrix using the "amorphous" module. Four types of structures were considered for this study. In the first type, long CNTs were embedded in the MG matrix. The second case consisted of

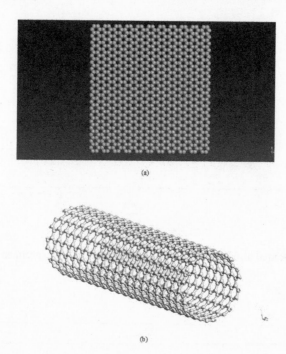

FIGURE 3.4 Initial configuration of (a) graphene sample having dimensions of 50 × 50 Å and (b) (10,10) armchair CNT.

embedding short CNTs in the MG matrix. In the third case, a long graphene sheet was used as reinforcement whereas, in the last case, a short graphene sheet was inserted in the MG matrix. Figure 3.5 shows the MG ($Cu_{64}Zr_{36}$) reinforced with (a) long CNT and (b) short CNT. In a similar manner, MG ($Cu_{64}Zr_{36}$) reinforced with (a) long graphene and (b) short graphene is shown in Figure 3.6. The effect of increasing the volume fraction of the reinforcement (CNT or graphene) and the effect of increasing the strain were studied. The aim of the study was to find which reinforcement gives better mechanical properties.

MD simulations were conducted with canonical (NVT) ensemble at a temperature of 300 K. Initial velocities were taken as random. Temperature was controlled by a Nose-Hoover thermostat [35]. Before applying the tensile loading, the MD systems were equilibrated for 30,000 time steps. For all the simulations, periodic boundary conditions (PBCs) were applied to the simulation box so as to make it as a representative volume element imitating the bulk phase of the nanocomposite. The elastic moduli were calculated by directly computing the average mechanical forces developed between carbon atoms in the nanotube. The effective elastic moduli can be calculated directly from the virial theorem given by Swenson [36] in which the expression of the stress tensor in a macroscopic system is given as the function of atom coordinates and inter-atomic forces. The method provides a continuum measure of the internal mechanical interactions between atoms. In an atomistic calculation, the internal stress tensor can be obtained using the so-called virial expression given by Swenson [36] as follows.

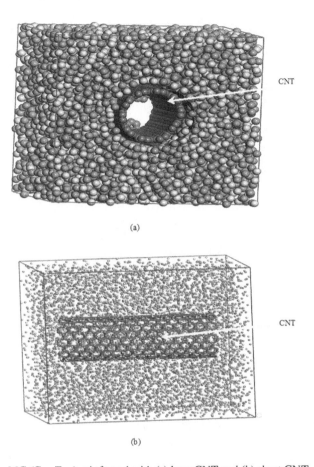

(a)

(b)

FIGURE 3.5 MG ($Cu_{64}Zr_{36}$) reinforced with (a) long CNT and (b) short CNT.

$$\sigma = -\frac{1}{V_0}\left[\left(\sum_{i=1}^{n} m_i(v_i v_i^T)\right) + \left(\sum_{i<j} r_{ij} f_{ij}^T\right)\right] \tag{3.1}$$

where index i runs over all particles 1 through N; m_i, v_i, and f_i denote the mass, velocity, and force acting on particle i; and V_0 denotes the (un-deformed) system volume. The application of stress to a body results in a change in the relative positions of particles within the body, expressed quantitatively via the strain tensor:

$$\varepsilon_{ij} = \begin{bmatrix} \varepsilon_{11} & \varepsilon_{12} & \varepsilon_{13} \\ \varepsilon_{21} & \varepsilon_{22} & \varepsilon_{23} \\ \varepsilon_{31} & \varepsilon_{32} & \varepsilon_{33} \end{bmatrix} \tag{3.2}$$

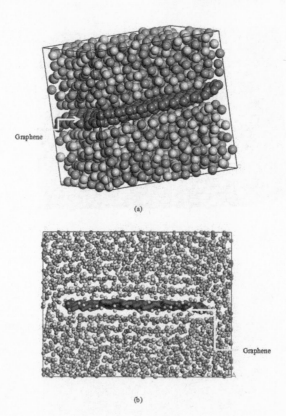

(a)

(b)

FIGURE 3.6 MG ($Cu_{64}Zr_{36}$) reinforced with (a) long graphene and (b) short graphene.

The elastic stiffness coefficients, relating the various components of stress and strain, are defined by:

$$C_{lmnk} = \frac{\partial \sigma_{lm}}{\partial \varepsilon_{nk}}\bigg|_{T, z_{nk}} = \frac{1}{V_0} \frac{\partial^2 A}{\partial \varepsilon_{lm} \partial \varepsilon_{nk}}\bigg|_{T, z_{lm}, z_{nk}} \quad (3.3)$$

where A denotes the Helmholtz free energy. For small deformations, the relationship between the stresses and strains may be expressed in terms of a generalized Hooke's law:

$$\sigma_{lm} = C_{lmnk} \varepsilon_{nk} \quad (3.4)$$

To calculate the axial Young's modulus (E_{11}), the atoms are displaced by $u_1 = \varepsilon_{11}^0$. The average strain and stresses are:

$$\varepsilon_{11} = \varepsilon_{11}^0$$

$$\sigma_{11} \neq 0, \text{ any other } \sigma_{ij} = 0 \quad (3.5)$$

Through MD simulation, the longitudinal elastic modulus, E_{11}, can be calculated as:

$$E_{11} = \frac{\sigma_{11}}{\varepsilon_{11}^0} \tag{3.6}$$

In other simulation runs, the load was applied either in the transverse or shear direction. On the lateral surfaces of the simulation box, stress-free conditions were imposed in order to satisfy the simple tension condition. The internal stress tensor is then obtained from the analytically calculated virial and used to obtain estimates of the six columns of the elastic stiffness coefficients matrix. A detailed explanation about the procedure for finding the elastic constants using MD can be found in [37]. The engineering constants may be calculated from elastic constants using the relations given by Christensen [38].

$$E_{33} = E_{22} = C_{22} + \frac{C_{12}^2\left(-C_{22}+C_{23}\right)+C_{23}\left(-C_{11}C_{23}+C_{12}^2\right)}{C_{11}C_{22}-C_{12}^2}$$

$$\nu_{12} = \nu_{13} = \frac{C_{12}}{C_{22}+C_{23}}$$

$$G_{23} = \frac{1}{2}\left(C_{22}-C_{23}\right) \tag{3.7}$$

$$K_{23} = \frac{1}{2}(C_{22}+C_{23})$$

3.2.3 THERMAL CONDUCTIVITY

3.2.3.1 Thermal Conductivity from MD

The thermal conductivity of a material is dominated by atomic vibrations, or phonons, and the conduction by electrons is generally negligible for insulating materials. For CNT/metallic glass nanocomposites, the thermal conductivity depends on several factors, including content, aspect ratio, dispersion of CNTs, and their interfacial interactions with the polymer matrix. The excellent thermal properties of CNTs, such as high thermal conductivity and good thermal stability, led to the expectation that CNTs could make useful functional fillers to rectify the thermal properties of metallic glass composites. Contrary to this expectation, the enhancements of thermal properties due to the incorporation of CNTs in metallic glass composites have not been quite remarkable, with the exception of a few isolated cases.

Many efforts have been devoted to employing CNTs as thermal conducting filler in metallic glass composites, and some enhancements have indeed been observed. Wang et al. [39] prepared Fe-based bulk amorphous alloy with low thermal conductivity by doping $Co_{65}Cr_{15}Zr_{10}W_{10}$ alloy in $Fe_{41}Co_7Cr_{15}Mo_{14}C_{15}B_6Y_2$ amorphous alloy.

The thermal conductivity in the range of 25°C–150°C and the electrical resistivity of Fe-based bulk amorphous alloy at room temperature were tested by a laser flash thermal conductivity tester and a four probe resistance tester respectively. The effect of the doping of $Co_{65}Cr_{15}Zr_{10}W_{10}$ amorphous alloy on the electrical resistivity and thermal conductivity and the relationship between the temperature and the thermal conductivity were investigated respectively. The results showed that, in the range of 25°C–150°C, the thermal conductivity of the Fe-based bulk amorphous alloy increased linearly with the increase of temperature. The thermal conductivity and the electrical resistivity of the Fe-based bulk amorphous alloy were decreased by doping $Co_{65}Cr_{15}Zr_{10}W_{10}$ alloy. When the doping content was higher than 10 at%, the thermal conductivity tended to be stable. Doping mainly results in the change of phonon thermal conductivity at room temperature, and the effect of doping on the electron thermal conductivity was very small. A formula for calculating the thermal conductivity of the Fe-based bulk amorphous alloy at room temperature was presented by using crystallization Tx. The relationship between phonon thermal conductivity and Tx was in linearity with $Y = 0.06756X - 37.31568$; the relationship between total thermal conductivity and Tx was in linearity with $Y = 0.06228X - 30.58814$. Liu et al. [40] focused on the role of 0-D defects (dopants, vacancies, interstitials, and antisites), 1-D defects (dislocations), 2-D defects (grain boundaries), and 3-D defects (nanoinclusions) in a benchmark thermoelectric material, Bi_2Te_3. The results gave new insights into developing higher-performance thermoelectric materials via defect engineering.

In this study, the imposed flux method was used to find thermal conductivity. A script was written in Materials Studio 7.0 using the available scripting option. The number of layers in which the direction of flux is divided was fixed at 40. Increasing the number of layers can increase the accuracy of the gradients, but too many layers would have led to large fluctuations in the layer temperatures. Two types of exchange method were used in the study. First, the variable method exchanged a variable energy between one object in the hot layer and one in the cold layer. The fixed method exchanged a constant energy between all hot objects in the hot layer and all objects in the cold layer. The amount of energy to exchange in each step when using the fixed exchange type was taken as 1 kcal/mol. The flux was determined by the ratio of exchange energy and number of steps. The number of exchanges during the equilibration stage was taken as 500. During the equilibration stage, a thermostat (NVT) acted on the system. The number of exchanges during the production stage was equal to 1000. The production stage was carried out at constant energy (NVE). A time step of 1 fs was used in the simulation. The number of time steps in between two exchanges was fixed at 100. Decreasing the number of steps led to higher fluxes and increased the temperature gradient. Too small values were avoided as these introduced non-linear effects and might impact performance.

To find thermal conductivity, MD simulations were performed with SWCNT (or graphene) volume fractions varying from $V_f = 0$–0.16 and the aspect ratio (in the case of CNTs) was kept fixed at $l/d = 10$. The results obtained from MD have been compared with analytical models for validation of results. These models have been discussed below.

3.2.3.2   Models to Calculate Thermal Conductivity

Many theoretical and semitheoretical models are available to represent the effective thermal conductivity of conventional polymer composites in which large-size fillers have been dispersed in a polymer matrix. Simple models, such as the series model, give the lower bound whereas the parallel model (rule of mixture) gives the upper bound of the thermal conductivity of a nanocomposite. As might be expected, experimental observations suggest that real values for nanocomposites fall somewhere in between these two limits.

1. *Parallel and series models:* The series model and the parallel model both assume that each phase contributes independently to the overall thermal resistance and conductance, respectively, and they assume a perfect interface between any two phases in contact. The series model applies readily to the thermal conductivity of a laminated composite along the stacking direction. However, it typically gives an underestimation of thermal conductivity due to the presumably complete localization of the contribution from the fibers embedded in the matrix, neglecting the interaction among the fillers. Therefore, the series model gives the lowest bound for the thermal conductivity of composites. The series model of thermal conductivity is given by;

$$k = (k_m \times k_f) / (k_f \times (1 - v_f) + k_m \times v_f) \tag{3.8}$$

where:

k = thermal conductivity of composite

k_m = thermal conductivity of Cu-Zr metallic glass = 12 W/m/K [41]

k_f (for SWCNT) = thermal conductivity of (10,10) armchair SWCNT = 9750 W/m/K [42]

k_f (for graphene) = thermal conductivity of graphene = 3000 W/m/K [43]

v_f = SWCNT/graphene volume fraction

In comparison, the parallel model predicts the thermal conductivity of conventional continuous fiber-reinforced composites along the fiber alignment direction. The rule of mixture implicitly assumes perfect contact between fibers. However, it gives a large overestimation of thermal conductivity and gives an upper bound for the thermal conductivity of composites. It is worth pointing out that thermal conductivity measurement results of composites should always fall between the predictions by the series model (lower bound) and the parallel model (upper bound) except for the cases where interfacial phonon scattering in nanolaminates can yield even lower thermal conductivity than the lower bound by the series model.

For the parallel model, thermal conductivity is given as:

$$k = (1 - v_f) \times k_m + v_f \times k_f \tag{3.9}$$

where the symbols have the same meaning as previously described in the series model.

2. *Maxwell-Garnett model:* The problem of determining the effective transport properties of multiphase materials dates back to Maxwell. Based on the continuity of potential and electric current at the interface, and on the assumption that the interactions among the fibers are negligible, which means the disseminated fibers are located far enough from each other, Maxwell derived an analytical formula for the effective specific resistance (K) of "a compound medium consisting of a substance of specific resistance K_2, in which are disseminated small spheres of specific resistance K_1, the ratio of the volume of all the small spheres to that of the whole being p." When transformed to the thermal conductivity (k) discussed here, the model gives:

$$k = k_m + \left[3 \times v_f \times (k_f - k_m) / (2 \times k_m + k_f - v_f \times (k_f - k_m)) \right] \quad (3.10)$$

Equation (3.10) gives satisfactory results for composites with (i) very low v_f, (ii) good dispersion, and (iii) no interfacial thermal resistance. It is also named the Maxwell-Garnett (MG) equation.

3. *Lewis-Nielsen model:* In this model, conductivity becomes the analogue of the stiffness, or elastic shear modulus, and the disturbance of the flux field becomes analogous to the disturbance of the stress field by the dispersed filler. Starting from Halpin-Tsai [44] equations, which are widely used in micromechanics, Nielsen applied a modified equation, the Nielsen-Lewis equation, to the modeling of thermal conductivity:

$$k = k_m \times \left[(1 + A \times B \times v_f) / (1 - B \times \psi \times v_f) \right] \quad (3.11)$$

where

$$A = k_E - 1,$$

k_E is the generalized Einstein coefficient, which depends primarily upon the shape of the fillers and how they are oriented with respect to the direction of the heat flow. For our case,

$$A = 2l/d = 10,$$

$$B = \left[(k_f / k_m) - 1 \right] / \left[(k_f / k_m) + A \right]$$

$$\psi = 1 + 1.775 \times v_f$$

Although Nielsen's model is a semi-empirical model, significant improvements in the model should be appreciated. The shape effect and, to some extent, the orientation effect are both taken into account. Reduced filler loading (ψ) accounts for the maximum packing density of the fillers with a specific shape and size distribution and is unique for this model. In comparison, most of the theoretical equations assume uniform changes of filler loading up to the point where the dispersed phase makes up the complete system, which is not realistic. The Nielsen-Lewis equation gives a higher prediction than the Maxwell-Garnett equation mainly because of the reduced filler loading. However, we should note that the model gives too high a prediction at high filler loading. In addition, interfacial thermal resistance is not considered in this model.

4. *Hamilton-Crosser model:* The Hamilton and Crosser model [45] is an extension of Maxwell's theory, accounting for the nonsphericity of fillers through the use of a shape factor, n, defined as $n = 3/\psi$ with ψ being the particle sphericity. The sphericity ($\psi = A_e/A$) of a particle is defined as the ratio of the surface area (A_e) of the equivalent sphere having the same volume to the actual surface area (A) of the nonspherical fiber. The effective thermal conductivity of the Hamilton and Crosser model is given by:

$$k = k_m \left[\frac{k_f + 5k_m - 5v_f \times \left(k_m - k_f\right)}{k_f + 5k_m + v_f \times \left(k_m - k_f\right)} \right] \tag{3.12}$$

Since neither interfacial resistance nor fiber-fiber interaction was taken into account, fiber size was found to have no influence on the effective thermal conductivity of the composite in this model.

5. *Deng model:* For estimating the mechanical and thermal properties of inclusion-in-matrix composites, even though for those with high concentrated inclusions so that interactions among inclusions must be considered, some mature methods have been established within the framework of micromechanics. For CNT composites with low loadings of randomly oriented straight CNTs of average length L and diameter d, an analytical estimate for the effective thermal conductivities, k_e, of the CNT composites is given by Eq. (3.13).

$$\frac{k_e}{k_m} = 1 + \frac{\dfrac{v_f}{3}}{\dfrac{k_m}{k_f} + H} \tag{3.13}$$

where

$$H = \frac{1}{p^2 - 1} \left[\frac{p}{\sqrt{p^2 - 1}} \ln\left(p + \sqrt{p^2 - 1}\right) - 1 \right],$$

$$p = l/d.$$

3.3   RESULTS AND DISCUSSION

To validate the computational model, first, the perfect SLGS sample with the dimensions of 50 Å × 50 Å was analyzed through tensile loading. Figure 3.4a depicts the initial configuration of the sample. Figure 3.7 shows the stress-strain curve for pure graphene. The Young's modulus of SLGS obtained was 1.02 TPa. As shown in Table 3.1, this value is in agreement with those available in the literature. In addition, the fracture strain of pure graphene was obtained as 18.5%. This is in accordance with the numerical [47] and experimental studies [48].

In the second step, the $Cu_{64}Zr_{36}$ MG was analyzed. The size of the computational cell was 50 Å × 50 Å × 50 Å, and it contained a total of 8190 atoms (see Figure 3.1). The Young's modulus of MG was found to be 60.3 GPa at a temperature of 300 K, which is in agreement with the value of 59.4 GPa reported by Deng et al. [49]. They examined the influence of CNTs on the mechanical characteristics of amorphous metallic glasses by using MD simulations. Additionally, the yield strength and the corresponding strain of MG were obtained as 2.4 GPa and 4.5%, respectively. This strength is defined as the stress at which the first plastic deformations appeared following a significant reduction in the measured stress. Table 3.2 summarizes the

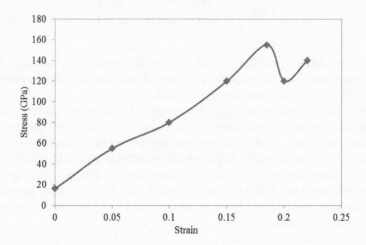

FIGURE 3.7   Stress-strain plot of pure graphene.

TABLE 3.1

Comparison of Young's Modulus of Graphene Obtained from the Present Study with Those Available in the Literature

Study	Simulation Approach	Young's Modulus (TPa)
Gupta et al. [42]	MD (Brenner potential)	1.27
Pei [43]	MD (AIREBO)	0.83
Liu et al. [46]	DFT	1.05
Present study	MD (COMPASS)	1.02

TABLE 3.2

Mechanical Properties of the MG Specimen

E (GPa)	σ_y (GPa)	ε_y
60.3	2.4	0.045

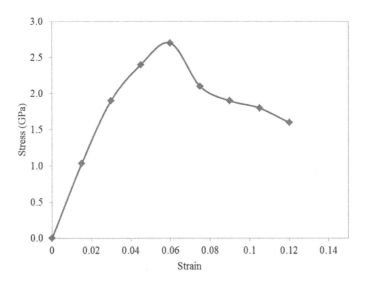

FIGURE 3.8 Stress-strain plot of the MG specimen.

results for mechanical properties of the MG the specimen. The stress-strain curve of the MG specimen in Figure 3.8 shows that the system undergoes considerable yielding under tensile loading, resulting in a significant reduction in stress.

Next, we compared the reinforcement effect of SLGS and CNT in improving the mechanical properties of MG. Both SLGS/MG and CNT/MG nanocomposites were analyzed for their elastic constants and mechanical properties. The focus was to find which nanofiller out of SLGS and CNT can significantly improve the mechanical properties of the MG matrix. For a better comparison, the nanofiller V_f and the implemented PBCs were the same for both composites. For modeling the CNT/MG case, a (10,10) armchair SWCNT with a diameter and length of 13.56 Å and 49.18 Å, respectively, was chosen. Two types of CNTs were considered, one long and the other one short. In long CNT/MG composites, the two ends of CNT touched the end faces of the periodic cell as shown in Figure 3.5a whereas, in short CNT/MG composites, the CNT was fully embedded inside the MG matrix as shown in Figure 3.5b. The CNT volume fraction (V_f) was varied from 4% to 16% for both short and long CNT–reinforced MG composites. Figures 3.9a and b show the stress-strain behavior of CNT/MG nanocomposites. The strain was applied along the direction of the CNTs. Comparing Figures 3.9a and b, it can be concluded that the long CNTs significantly improve the mechanical properties of the metallic material. The tensile strength and elastic modulus increased considerably. Also, the area under the curve

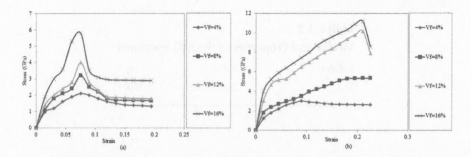

FIGURE 3.9 Stress-strain curve for (a) short CNT ($l/d = 4$) and (b) long CNT–reinforced MG composites, showing the effect of increase in volume fraction (V_f) on tensile stress-strain behavior.

for long CNT/MG composite was found to be larger in comparison to short CNT/MG composite. Thus, it was concluded that the long CNT/MG composites have higher toughness than short CNT/MG composites. The mechanical properties of short and long CNT/MG composites for $V_f = 12\%$ have been summarized in Table 3.3. It can be inferred that the elastic modulus of long CNT/MG composite increased approximately two times compared to pure MG. Also, the yield strength and the corresponding strain were found to increase by approximately four and three times, respectively, compared to the pure MG.

From Figure 3.9b, it can be concluded that the short CNT/MG composites have lower elastic modulus, yield strength, and yield strain in comparison to long CNT/MG composites. This underlines the inefficacious load transfer from MG to CNTs owing to the small aspect ratio ($l/d = 4$) of CNTs. Furthermore, from Table 3.3, it can be observed that the elastic modulus of short CNT/MG composite increased only by 15.4%. This increase in modulus could be attributed to the efficient reinforcement in the elastic region of the composite. In the plastic region, no enhancement in the the elastic modulus was observed for short CNT/MG composites. Figures 3.9a and b show the effect of increase in V_f of CNTs on the stress-strain behavior of CNT/MG composites. An increase in the mechanical properties was observed by increasing the V_f from 4% to 16%. The CNT V_f was increased by changing the matrix cross-section only, keeping the structure of CNT the same. The V_f of 12% was used for comparing all the simulation results.

Similar to the two types of CNTs considered above, two types of graphene were also studied, one long and the other one short. In long graphene/MG composites,

TABLE 3.3

Mechanical Properties of Short and Long CNT–reinforced MG Composites at a Volume Fraction of 12%

	E (GPa)	σ_y	ε_y
MG	60.3	2.4	0.05
Short CNT-MG	69.6	2.8	0.06
Long CNT-MG	115.5	10.2	0.21

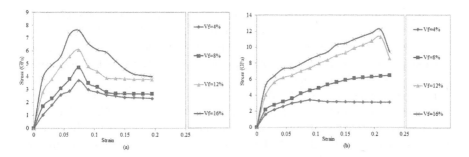

FIGURE 3.10 Stress-strain curve for (a) short and (b) long graphene–reinforced MG composites, showing the effect of increase in volume fraction (V_f) on tensile stress-strain behavior.

the two edges of graphene sheet touched the end faces of the periodic cell as shown in Figure 3.6a whereas, in short graphene/MG composites, the graphene was fully embedded inside the MG matrix as shown in Figure 3.6b. The graphene volume fraction (V_f) was also varied from 4% to 16% for both short and long graphene–reinforced MG composites. Figures 3.10a and b show the stress-strain behavior of graphene/MG nanocomposites. The strain was applied along the same direction as that for CNTs.

Comparing Figures 3.10a and b, it can be concluded that the long graphene significantly improves the mechanical properties of the metallic material. The tensile strength and elastic modulus increased considerably. Similar to the case of CNT/MG composites, the area under the curve for long graphene/MG composite was found to be larger in comparison to short graphene/MG composite. Thus, it was concluded that the long graphene/MG composites have higher toughness than short graphene/MG composites. The mechanical properties of short and long graphene/MG composites for $V_f = 12\%$ have been summarized in Table 3.4. It can be inferred that the elastic modulus of long graphene/MG composite increased approximately 2.15 times compared to pure MG. Also, the yield strength and the corresponding strain were found to increase by approximately five times each compared to the pure MG.

From Figure 3.10b, it can be concluded that the short graphene/MG composites have lower elastic modulus, yield strength, and yield strain in comparison to long graphene/MG composites. This again shows the inefficient load transfer from MG to graphene. Furthermore, from Table 3.4, it can be observed that the elastic modulus of short graphene/MG composite increased only by 25%. Figures 3.10a and b show

TABLE 3.4

Mechanical Properties of Short and Long Graphene Reinforced MG Composites at a Volume Fraction of 12%

	E (GPa)	σ_y	ε_y
MG	60.3	2.4	0.05
Short graphene-MG	75.4	2.9	0.07
Long graphene-MG	129.4	11.3	0.23

the effect of increase in V_f of graphene on the stress-strain behavior of graphene/MG composites. An increase in the mechanical properties was observed by increasing the V_f from 4% to 16%. The graphene V_f was increased by changing the matrix cross-section only, keeping the structure of graphene the same. The V_f of 12% was used for comparing all the simulation results.

Comparing Figures 3.9b and 3.10b, it can be concluded that long graphene/MG composites have better mechanical properties in comparison to the long CNT/MG composites. For the same value of strain (0.21), the maximum stress in graphene/MG composite was found to be approximately 12 GPa whereas, in CNT/MG composite, the stress was approximately 11 GPa. The percentage increase in elastic modulus with an increase in V_f from 0% to 12% for long graphene/MG composite was 114.6% and, for long CNT/MG composite, the percentage increase was approximately 91.54% compared to pure MG. Also, the percentage increase in yield strength and the corresponding strain for long graphene/CNT composite were found to be 371% and 360%, respectively. For long CNT/MG composites, the yield strength and strain were found to increase by 325% and 320%, respectively. The results show that long graphene/MG composites are better reinforcement for MG in comparison to long CNT/MG composites.

Figures 3.11 through 3.14 show the snapshots of deformation at different strain values. Figure 3.11 show that when the strain was increased to 0.15, the

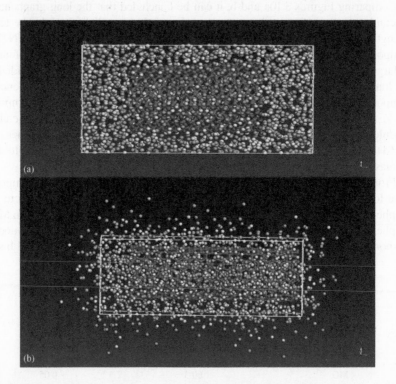

FIGURE 3.11   Snapshots of deformation of short CNT–reinforced MG composites at different strain values: (a) strain = 0 and (b) strain = 0.15.

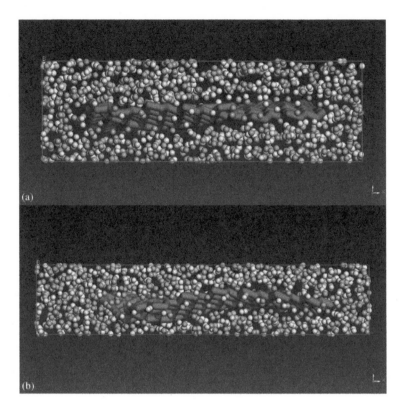

FIGURE 3.12   Snapshots of deformation of short graphene–reinforced MG composites at different strain values: (a) strain = 0 and (b) strain = 0.15.

short CNT tried to move toward the right side of the figure. Since the CNT was embedded in the MG matrix, it was prevented from being pulled out of the specimen. The length of the short CNT was found to increase considerably but no debonding was observed till the strain of 0.15. From Figure 3.12, it was observed that the short graphene sheet shifted by a small amount within the MG matrix, but its shape remained undistorted. The short nanofillers increase the elastic modulus only when the strain applied is within the elastic limit. Figure 3.13 shows the bending of long CNT. This bending effect could be due to the applied strain being equal to or greater than the ultimate strain of the CNT. No such deformation was observed for the short CNT. But due to the large surface area of long CNTs in comparison to the short ones, their elastic moduli were found to be greater. Figure 3.14 shows the snapshots of deformation of long graphene/MG composite. Contrary to the bending observed in long CNT/MG composites, the graphene/MG composite showed no such deformation. Graphene-reinforced MG composites were thus found to have better mechanical properties in comparison to CNT/MG composites.

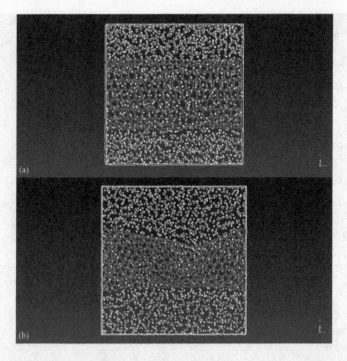

FIGURE 3.13 Snapshots of deformation of long CNT–reinforced MG composites at different strain values: (a) strain = 0 and (b) strain = 0.21.

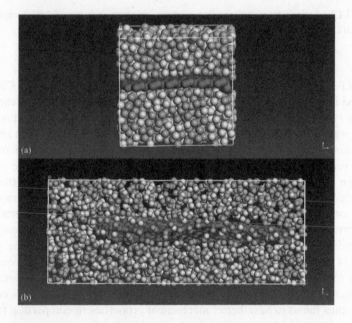

FIGURE 3.14 Snapshots of deformation of long graphene–reinforced MG composites at different strain values: (a) strain = 0 and (b) strain = 0.21.

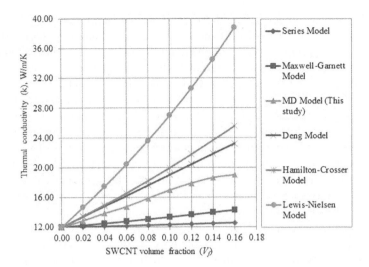

FIGURE 3.15 Comparison of MD results of thermal conductivity for (10,10) armchair SWCNT–reinforced MG composite with other models with a fixed aspect ratio ($l/d = 10$) and varying volume fraction.

Figure 3.15 shows the comparison of MD results of thermal conductivity for armchair SWCNT–reinforced metallic glass composite with other models with a fixed aspect ratio ($l/d = 10$). Figure 3.16 shows the comparison of MD results of thermal conductivity for graphene-reinforced metallic glass composite with other models. Thermal conductivity shows an increasing trend with an increase in

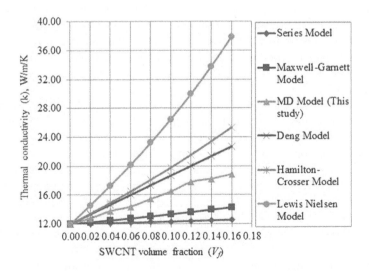

FIGURE 3.16 Comparison of MD results of thermal conductivity for graphene-reinforced MG composite with other models and varying volume fraction.

SWCNT/graphene volume fraction. Comparing Figures 3.15 and 3.16, it could be concluded that SWCNT-reinforced metallic glass composites have higher thermal conductivity in comparison to graphene-reinforced metallic glass composites. The series model and the parallel model both assume that each phase contributes independently to the overall thermal resistance and conductance, respectively, and they assume a perfect interface between any two phases in contact. However, it typically gives an underestimation for a particulate composite due to the presumably complete localization of the contribution from the particles embedded in the matrix that is neglecting the interaction among the fillers. Therefore, the series model gives the lowest bound for the thermal conductivity of composites. In comparison, the parallel model predicts the thermal conductivity of conventional continuous fiber-reinforced composites along the fiber alignment direction. For composites with fibrous inclusions, the rule of mixture implicitly assumes perfect contact between particles in a fully percolating network. However, it gives a large overestimation of thermal conductivity for other types of composites and gives an upper bound for the thermal conductivity of composites. It is worth pointing out that thermal conductivity measurement results of composites should always fall between the predictions by the series model (lower bound) and the parallel model (upper bound) except for the cases where interfacial phonon scattering in nanolaminates can yield even lower thermal conductivity than the lower bound by the series model.

Based on the continuity of potential and electric current at the interface, and on the assumption that the interactions among the spherical fillers are negligible, which means the disseminated small spheres are located far enough from each other, Maxwell derived an analytical formula for the effective specific resistance (K) of "a compound medium consisting of a substance of specific resistance K_2, in which are disseminated small spheres of specific resistance K_1, the ratio of the volume of all the small spheres to that of the whole being p." This model gives satisfactory results for composites with: (i) spherical inclusions, (ii) very low V_f, (iii) good dispersion, and (iv) no interfacial thermal resistance. Since neither interfacial resistance nor particle-particle interaction was taken into account in the Hamilton-Crosser model, fiber size was found to have no influence on the effective thermal conductivity of the composite in this model.

Although Nielsen's model is a semi-empirical model, at least three important improvements in the model should be appreciated. First, the shape effect and, to some extent, the orientation effect are both taken into account. Second, reduced filler loading accounts for the maximum packing density of the fillers with a specific shape and size distribution and is unique for this model. In comparison, most of the theoretical equations assume uniform changes of filler loading up to the point where the dispersed phase makes up the complete system, which is not realistic. Third, the earliest definition of "effective unit" is reflected in the discussion on aggregates of spheres. Table 3.5 shows a comparison of MD results for thermal conductivity of armchair SWCNT–reinforced metallic glass composite with other models. It could be inferred that at all SWCNT volume fractions, MD results agreed well with the Deng model rather than with the Maxwell-Garnett model.

TABLE 3.5

Comparison of MD Results for Thermal Conductivity of (10,10) Armchair SWCNT–Reinforced Metallic Glass Composite with Other Models

SWCNT V_f (in %)	Percentage Difference in Thermal Conductivity Between	
	Maxwell-Garnett Model and MD	MD and Deng Model
0	0	0
2	4.785	4.045
4	9.752	6.447
6	13.400	9.050
8	17.819	9.875
10	21.440	10.736
12	23.789	12.361
14	25.117	14.602
16	24.869	18.123

3.4 CONCLUSIONS

MD simulations were performed to investigate the effect of CNT and graphene reinforcements on the mechanical properties of amorphous MG. It was concluded that the CNTs can significantly enhance the elastic modulus, yield strength, and yield strain of amorphous MG, but graphene was found to be a better reinforcement. But when compared on the basis of thermal conductivity, SWCNT proved to be a better reinforcement in comparison to graphene. Some of the main conclusions that can be drawn based on the research conducted are highlighted below.

1. Long CNTs considerably improve the mechanical properties of MG whereas the short CNTs proved to be inefficient in improving the properties. This could be due to the nonbonded interactions between the MG and reinforcement interface and also due to the poor load bearing capacity of the unreinforced portion of MG in the case of short CNT/MG composites.
2. Long CNT- and graphene-reinforced MG specimens have high toughness in comparison to the short CNT and graphene MG composites as was observed from Figures 3.9 and 3.10.
3. The tensile strength of long graphene/MG composite was 12 GPa whereas, for long CNT/MG composite, the tensile strength was found to be 11 GPa. The percentage increase in elastic modulus with an increase in V_f (from 0% to 12%) of long graphene/MG composite was found to be 114.6%, and for long CNT/MG composite the percentage increase was approximately 91.54% compared to pure MG.

4. The percentage increase in yield strength and the corresponding strain for long graphene/CNT composite were found to be 371% and 360%, respectively. For long CNT/MG composites, the yield strength and strain were found to increase by 325% and 320%, respectively. The results showed that the long graphene/MG composites are better reinforcement for MG in comparison to the long CNT/MG composites.

5. Thermal conductivity values for SWCNT-reinforced MG composites were found to be higher in comparison to graphene-reinforced MG composites for the same V_f.

The results of the study will help the researchers in the design and development of such MG-based nanocomposites that have high strength and ductility as well as high thermal conductivity.

ACKNOWLEDGEMENT

The authors state that this research has not been funded by any source and has received no funds from any institute or organization.

REFERENCES

1. Endo, M., Natsuki, T., Tantrakarn, K. (2004), "Effects of carbon nanotube structures on mechanical properties," *Applied Physics A: Materials Science & Processing*, Vol. 79, No. 1, pp. 117–124.

2. Liu, X., Metcalf, T.H., Robinson, J.T., Houston, B.H., Scarpa, F. (2012), "Shear modulus of monolayer graphene prepared by chemical vapor deposition," *Nano Letters*, Vol. 12, No. 2, pp. 1013–1017.

3. Ghosh, S., Calizo, I., Teweldebrhan, D., Pokatilov, E.P., Nika, D.L., Balandin, A.A., Bao, W., Miao, F., Lau, C.N. (2008), "Extremely high thermal conductivity of graphene: Prospects for thermal management applications in nanoelectronic circuits," *Applied Physics Letters*, Vol. 92, No. 15, pp. 151911–151913.

4. Ghosh, S., Bao, W., Nika, D.L., Subrina, S., Pokatilov, E.P., Lau, C.N., Balandin, A.A. (2010), "Dimensional crossover of thermal transport in few-layer graphene," *Nature Materials*, Vol. 9, No. 7, pp. 555–558.

5. Yu, M.F., Lourie, O., Dyer, M., Moloni, K., Kelly, T.F., Ruoff, R.S. (2000), "Strength and breaking mechanism of multi walled carbon nanotubes under tensile load," *Science*, Vol. 287, No. 5453, pp. 637–640.

6. Chen, H.S., Ferris, S.D., Gyorgy, E.M., Leamy, H.J., Sherwood, R.C. (1975), "Field heat treatment of ferromagnetic metallic glasses," *Applied Physics Letters*, Vol. 26, No. 7, pp. 405–406.

7. Chen, H.S. (1975), "Glassy metals," *Reports on Progress in Physics*, Vol. 43, No. 4, pp. 353–432.

8. Inoue, A., Zhang, T., Masumoto, T. (1989), "Al-La-Ni amorphous alloys with a wide super-cooled liquid region," *Materials Transactions JIM*, Vol. 30, No. 12, pp. 965–972.

9. Inoue, A., Zhang, T., Nishiyama, N., Ohba, K., Masumoto, T. (1993), "Preparation of 16 mm diameter rod of amorphous $Zr_{65}Al_{7.5}Ni_{10}Cu_{17.5}$ alloy," *Materials Transactions JIM*, Vol. 34, No. 12, pp. 1234–1237.

10. Nishiyama, N., Takenaka, K., Miura, H., Saidoh, N., Zeng, Y., Inoue, A. (2012), "The world's biggest glassy alloy ever made," *Intermetallics*, Vol. 30, No. 1, pp. 19–24.

11. Chen, M.W (2008), "Mechanical behaviour of metallic glasses: microscopic understanding of strength and ductility," *Annual Review of Materials Research*, Vol. 38, No. 1, pp. 445–469.
12. Kawamura, Y., Nakamura, T., Inoue, A., Masumoto, T. (1999), "High strain rate superplasticity due to Newtonian viscous flow in $La_{55}Al_{25}Ni_{20}$ metallic glass," *Materials Transactions JIM*, Vol. 40, No. 8, pp. 794–803.
13. Wang, W.H., Bian, Z., Pan, M.X., Zhang, Y. (2002), "Carbon nanotube reinforced $Zr_{52.5}Cu_{17.9}Ni_{14.6}Al_{10}Ti_5$ bulk metallic glass composites," *Applied Physics Letters*, Vol. 81, No. 25, pp. 4739–4741.
14. Mukai, T., Nieh, T.G., Kawamura, Y., Inoue, A., Higashi, K. (2002), "Dynamic response of a $Pd_{40}Ni_{40}P_{20}$ bulk metallic glass in tension," *Scripta Materialia*, Vol. 46, No. 1, pp. 43–47.
15. Ravichandran, G., Lu, J., Johnson, W.L. (2003), "Deformation behaviour of the $Zr_{41.2}Ti_{13.8}Cu_{12.5}Ni_{10}Be_{22.5}$ bulk metallic glass over a wide range of strain-rates and temperatures," *Acta Materialia*, Vol. 51, No. 12, pp. 3429–3443.
16. Johnson, W.L. and Samwer, K. (2005), "A universal criterion for plastic yielding of metallic glasses with a $(T/Tg)^{2/3}$ temperature dependence," *Physical Review Letters*, Vol. 95, No. 19, pp. 195501(1)–195501(4).
17. Li, M. and Li, Q.K. (2006), "Molecular dynamics simulation of intrinsic and extrinsic mechanical properties of amorphous metals," *Intermetallics*, Vol. 14, No. 8–9, pp. 1005–1010.
18. Sui, M.L., Guo, H., Yan, P.F., Wang, Y.B., Tan, J., Zhang, Z.F., Ma, E. (2007), "Tensile ductility and necking of metallic glass," *Nature Materials*, Vol. 6, No. 1, pp. 735–739.
19. Ramamurty, U., Dubach, A., Raghavan, R., Loffler, J.F., Michler, J. (2009), "Micropillar compression studies on a bulk metallic glass in different structural states," *Scripta Materialia*, Vol. 60, No. 7, pp. 567–570.
20. Jang, D. and Greer, J.R. (2010), "Transition from a strong-yet-brittle to a stronger-and-ductile state by size reduction of metallic glasses," *Nature Materials*, Vol. 9, No. 3, pp. 215–219.
21. Hosson, J.T.M.D., Kuzmin, O.V., Pei, Y.T., Chen, C.Q. (2012), "Intrinsic and extrinsic size effects in the deformation of metallic glass nanopillars," *Scripta Materialia*, Vol. 60, No. 3, pp. 889–898.
22. Misra, R.D.K., Li, S., Zhao, P., Gao, G., Bai, B. (2015), "Mechanical behaviour of carbon nanotube reinforced Mg-Cu-Gd-Ag bulk metallic glasses," *Materials Science & Engineering A*, Vol. 641, No. 1, pp. 116–122.
23. Gupta, M., Jayalakshmi, S., Sahu, S., Sankaranarayanan, S., Gupta, S. (2014), "Development of novel $Mg-Ni_{60}Nb_{40}$ amorphous particle reinforced composites with enhanced hardness and compressive response," *Materials and Design*, Vol. 53, No. 1, pp. 849–855.
24. Kalcher, C., Brink, T., Rohrer, J., Albe, K., Stukowski, A. (2017), "Interface-controlled creep in metallic glass composites," *Acta Materialia*, Vol. 141, No. 1, pp. 251–260.
25. Wang, Y., Li, M., Xu, J. (2017), "Mechanical properties of spinodal decomposed metallic glass composites," *Scripta Materialia*, Vol. 135, pp. 41–45.
26. Li, L., Li, J., Wang, J., Kou, H. (2017), "Tune the mechanical properties of Ti-based metallic glass composites by additions of nitrogen," *Materials Science and Engineering: A*, Vol. 694, pp. 93–97.
27. Lee, P.Y., Lin, Y.S., Hsu, C.F., Chen, J.Y., Cheng, Y.M. (2016), "Wear behavior of mechanically alloyed Ti-based bulk metallic glass composites containing carbon nanotubes," *Metals-Open Access Metallurgy Journal*, Vol. 6, No. 11, pp. 289–298.
28. Villapún, V.M., Esat, F., Bull, S., Dover, L.G., González, S. (2017), "Tuning the mechanical and antimicrobial performance of a Cu-based metallic glass composite through cooling rate control and annealing," *Materials (Basel)*, Vol. 10, No. 5, pp. 506–514.
29. Chen, S., Zhang, L., Fu, H., Zhang, H., Li, Z.K., Zhu, Z., Li, H., Zhang, H.W., Wang, A.M., Wang, Y.D. (2018), "Compressive mechanical properties and failure modes of Zr-based bulk metallic glass composites containing tungsten springs," *Materials and Design*, Vol. 160, pp. 652–660.

30. Hong, S.H., Kim, J.T., Park, J.M., Song, G., Wang, W.M., Kim, K.B. (2018), "Mechanical, deformation and fracture behaviors of bulk metallic glass composites reinforced by spherical B2 particles" *Progress in Natural Science: Materials International*, Vol. 28, No. 6, pp. 704–710.

31. Pan, J., Lin, Y., Zhang, J., Huang, W., Li, Y. (2019), "Effect of Ta particles on the fracture behavior of notched bulk metallic glass composites," *Intermetallics*, Vol. 106, pp. 1–6.

32. Cardinal, S., Pelletier, J.M., Xie, G.Q., Mercier, F., Dalmas, F. (2019), "Enhanced compressive plasticity in a Cu-Zr-Al–Based metallic glass composite," *Journal of Alloys and Compounds*, Vol. 782, pp. 59–68.

33. Sharma, S., Chandra, R., Kushwaha, P., Kumar, N. (2013), "Molecular dynamics simulation of polymer/CNT composites", *Acta Mechanica Solida Sinica*, Vol. 28, No. 4, pp. 409–419.

34. Sharma, S., Chandra, R., Kushwaha, P., Kumar, N. (2014), "Effect of Stone-Wales and vacancy defects on elastic moduli of carbon nanotubes and their composites using molecular dynamics simulation," *Computational Materials Science*, Vol. 86, pp. 1–8.

35. Hoover, W.G. (1985), "Canonical dynamics: Equilibrium phase-space distributions," *Physical Review A*, Vol. 31, No. 3, pp. 1695–1697.

36. Swenson, R.J. (1983), "Comments on virial theorems for bounded systems," *American Journal of Physics*, Vol. 51, No. 10, pp. 940–942.

37. Sharma, S., Chandra, R., Kushwaha, P., Kumar, N. (2016), "Mechanical properties of carbon nanofiber reinforced polymer composites-molecular dynamics approach," *Journal of the Minerals, Metals & Materials Society (TMS)*, Vol. 68, No. 6, pp. 1717–1727.

38. Christensen, R. (1991), *Mechanics of Composite Materials*, Krieger Publishing Company, Malbar, FL, pp. 74.

39. Wang, C.J., Chen, Q.J., Xia, H.X. (2017), "Effect of doping alloy system on thermal conductivity of FeCoCrMoCBY amorphous alloy," *Transactions of Materials and Heat Treatment*, Vol. 38, No. 3, pp. 16–20.

40. Liu, Y., Zhou, M., He, J. (2016), "Towards higher thermoelectric performance of Bi_2Te_3 via defect engineering", *Scripta Materialia*, Vol. 111, pp. 39–43.

41. Bykov, V.A., Kulikova, T.V., Yagodin, D.A., Fillippov, V.V., Shunyaev, K.Y. (2015), "Thermophysical and electrical properties of equiatomic CuZr alloy," *The Physics of Metals & Metallography*, Vol. 116, pp. 1067–1072.

42. Gupta, S., Dharamvir, K., Jindal, V.K. (2005), "Elastic moduli of single-walled carbon nanotubes and their ropes," *Physical Review B*, Vol. 72, No. 16, pp. 165428–165443.

43. Pei, Q.X., Zhang, Y.W., Shenoy, V.B. (2010), "Mechanical properties of methyl functionalized graphene: A molecular dynamics study," *Nanotechnology*, Vol. 21, No. 11, pp. 115709–115716.

44. Halpin, J.C. and Kardos, J.L. (1976), "The Halpin-Tsai equations: A review," *Polymer Engineering Science*, Vol. 16, pp. 344–352.

45. Hamilton, R.L. and Crosser, O.K. (1962), "Thermal conductivity of heterogeneous two-component systems," *Industrial and Engineering Chemistry Fundamentals*, Vol. 1, No. 3, pp. 187–191.

46. Liu, F., Ming, P., Li, J. (2007), "*Ab initio* calculation of ideal strength and phonon instability of graphene under tension," *Physical Review B*, Vol. 76, No. 6, pp. 064120–064126.

47. Ogata, S. and Shibutani, Y. (2003), "Ideal tensile strength and band gap of single-walled carbon nanotubes," *Physical Review B*, Vol. 68, No. 16, pp. 165409–165412.

48. Hone, J., Lee, C., Wei, X., Kysar, J.W. (2008), "Measurement of the elastic properties and intrinsic strength of monolayer graphene," *Science*, Vol. 321, No. 5887, pp. 385–388.

49. Deng, C., Rezaei, R., Shariati, M., Tavakoli-Anbaran, H. (2016), "Mechanical characteristics of CNT-reinforced metallic glass nanocomposites by molecular dynamics simulations," *Computational Materials Science*, Vol. 119, No. 1, pp. 19–26.

4 Predicting Thermal Conductivity of Metallic Glasses and Their Nanocomposites

Raja Sekhar Dondapati

CONTENTS

4.1 INTRODUCTION

The discovery of noncrystalline metallic structures reformed the field of material science and its processing [1]. Ever since this discovery, metallic glasses have considerably drawn the attention of researchers and industrialists in the past few decades. The unstructured distribution of metallic atoms gives rise to favorable characteristics, such as elevated toughness, strength, and wear and corrosion resistance [2]. Additionally, a schematic of the LDM process is shown in Figure 4.1. Also, the presence of atoms of less weight, such as calcium, magnesium, and titanium, takes into consideration the high strength to weight ratio for potential applications in transportation and aviation ventures. Moreover, metallic glasses are a promising candidate for biomedical inserts because of lower degradation rates, which give sufficient time for healing of tissues [3]. Extensive work has been conducted in the field of metallic glasses, which widens their applications in multidisciplinary fields. Lu et al. [4] extended the uses of metallic glasses as functionally graded materials, which removes their inherent issues, for instance, limited size estimations. Laser direct manufacturing (LDM) is a technique to produce structurally graded materials. In their investigation, a $Zr_{50}Ti_5Cu_{27}Ni_{10}Al_8$ (Zr_{50}) alloy was selected. Through finite element simulation, a Zr_{50} composite was produced. Zhang et al. [5] evaluated the impact of protons for bombarding $Ni_{62}Ta_{38}$ metallic glass and a metal W. The results exhibited that after proton illumination, $Ni_{62}Ta_{38}$ still kept up its amorphousness. The light coverage was 1.0×10^{18} particles/cm^2, where considerable roughness was found in metal W. The roughness of metal W increased under the impact of protons. Deb Nath [6] used molecular simulations to acquire thermal conductivity of various metallic glasses. Yielding and Young's modulus of metallic glasses, for example, $Zr_xCu_{90x}Al_{10}$, were procured, where the impact of strain rates and temperature were taken into account. Effects of strain rate on the yield nature and Young's modulus were analyzed. Lv and Chen [7] utilized the laser flicker and multifunction thermal conductivity instrument for obtaining the thermal conductivity of Fe- metallic glass with varying crystallization. Three stages of internal friction characteristics of $Fe_{45}Cr_{15}Mo_{14}C_{15}B_6Y_2Ni_3$ BMG were obtained, which were divided into three regions: superplastic region, crystallization, and high-strength region.

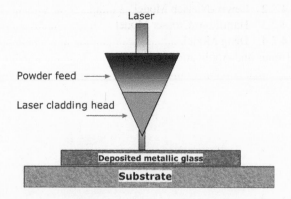

FIGURE 4.1 Schematic of the LDM process.

The thermal conductivity increased with the development of crystallization, as the crystallization and superplastic weakens. Presz and Kulik [8] demonstrated the microformation of metallic glass at elevated temperature under the influence of ultrasonic vibrations. Mistarihi et al. [9] proposed U-Mo dissipating fuel for test reactors and research purposes due to its consistent irradiation of U-Mo fuel in examination with metallic fuels. Regardless, U-Mo dissipating fuel demonstrates high growth of layers. Kbirou [10] proposed a molecular dynamics simulation for subnuclear components generation. In Co_2Al, basic qualities at medium-range order (MRO) and short-range order (SRO) were inspected by techniques, radical distribution function (RDF), Voronoi tessellation examination (VTA), and coordination number (CN). Lu et al. [11] gave a technique for mitigating the characteristic issues of metallic glass, such as brittleness and limited dimension, by using them as functionally graded material. To create such metallic glass, laser additive manufacturing (LAM) is a reliable technique [12].

Through limited component reproduction, an auxiliary reviewed Zr50 compound with a thickness bigger than 10 mm was created. Garnier and Boudenne [13] proposed an efficient way to improve the thermal conductivity of polymers with combined metallic particles. An objective of their work was to study and compare the effect of hollow metallic particles and full ones. The hollow particles proved to be better than full nanoparticles under increased thermal conductivity and lesser weight considerations. Boussatour et al. [14] evaluated the thermal conductivity in adaptable biosourced starting point polymers by using the 3-omega method. In their work, two biopolymers were considered: the cellulose palmitate (CP) and a polylactic destructive (PLA), which were created in their research facility. The approach relies upon thermal disturbance of the sample by utilizing a metal part. Thermal conductivities of order 0.19 and 0.30 W/m-K were obtained independently for PLA and CP films. Yamasaki et al. [15] measured the thermal diffusivity of formless, strong, and supercooled liquid in a $Zr_{55}Al_{10}Ni_5Cu_{30}$ (at%) BMG using a laser flash methodology. The thermal diffusivity and conductivity of the shapeless BMG were temperature-dependent, with minimal positive temperature coefficients. Diallo [16] estimated the thermal conductivity of aluminium nanoparticles in a nanoporous strong matrix in order to observe the effect of conductive filler on the thermal properties of a porous medium. A limited increase in thermal conductivity was observed in view of the possible mechanism involved: (1) decrease in porosity with enhanced solid-gas coupling (2) increase in contact zone between mix constituents. An exponential growth of thermal conductivity was observed as a result of coupling between the solid particles and the entrapped air. Zadorozhnyy [17] produced a bulk composite based on $Cu_{54}Pd_{28}P_{18}$ metallic glass and polytetrafluoroethylene (PTFE of around 1 mass %) by alloying with plasma sintering. Near the supercooled liquid zone temperature, the plasma sintering was executed. It was found that the thermal conductivity was dominantly increased due the precedent composite ($Cu_{54}Pd_{28}P_{18}$/PTFE). Henao et al. [18] performed the deposition of cold-sprayed metallic glass particles on metallic substrate, which was performed using numerical examination by opting ABAQUS code. Upon impact, the generation of stress and the thickening of substrate were examined, which showed the dependence on the properties of the substrate. Chen et al. [19] directed small-scale machining of $Zr_{55}Cu_{30}Al_{10}Ni_5$ mass metallic glass for respectability of surface, chip morphology, and wear cutting force upon precise cutting with a boron nitride instrument and a diamond tool. Thili et al. [20] inspected the thermal properties of Zr-based bulk metallic glass, which was noncrystalline in nature. Utilizing electric measurements and X-ray scattering, the effects of noncrystallization, electron transport, and molecular vibration on thermal properties were obtained. No interface enhancement phonon or electron scattering were observed. A moderate difference in the electrical conductivity was enhanced under a certain degree of crystallization. Their results showed that composites on loss crystallinity are a promising alternative for enhancing electron transport. Li et al. [21] concluded that additive manufacturing is a promising methodology for the production of BMG parts in the absence of size confinements. In systematic examinations joined with constrained segment entertainment, they found that scaled-down scale breaks in normally Fe-based metallic glass in the midst of selective laser melting (SLM) are actuated by thermal unsettling around littler scale pores. The results show that the period of high thickness withdrawals in second stages in the midst of SLM drastically decline thermal agitation by releasing strain and, in this manner, cover littler scale split advancement.

The adjustments to the surface morphology of metallic glasses $Fe_{68}Zr_7B_{25}$ and $Fe_{80}Si_{7.43}B_{12.57}$ under the radiation of He particles were studied [22]. The outcomes demonstrated that after the He^{2+} illumination, metallic glass $Fe_{68}Zr_7B_{25}$ still kept up amorphousness. The flowchart of the procedure for the calculation of thermophysical properties is shown in Figure 4.2. The measurement of metastable β-Mn type stage nanocrystals remained in amorphous nature at 4.0×10^{17} ions/cm^2 (19 dpa). Choy et al. [23] studied the thermal conductivity of three different metallic glasses with the help of thermal diffusivity, and it was found that the accuracy of calculated thermal conductivity was 8%. Wiedemann Franz law was used for the evaluation of the electronic component to the thermal conductivity and it was observed that it increases linearly with increasing temperature. It was also observed that the phonon thermal conductivity was dependent on temperature for weak and nonmagnetic alloys. Thermal conductivity cannot be easily determined due to the small thickness of the metallic glass samples and radiative loss.

Li et al. [24] reviewed that the microstructure of ferrous-based bulk metallic glasses can be transformed into amorphous composites from the structure with the reduction in cooling rate. Excellent soft magnetic characteristics, corrosion resistance, high strength, and hardness are exhibited by bulk metallic glasses. In addition, two main application areas of ferrous-based metallic glass are soft magnetic materials and coatings. Moreover, it also has high strength and elastic modulus and is adaptable for biomedical material applications.

Bulk metallic glasses have cleared the way for fundamental studies and technological advancement. Due to extraordinary glass-forming ability and the higher stability of these materials, they have facilitated the opportunity to study the glass transition, kinetics, thermodynamic properties, and transport properties. They have become ideal for structural engineering applications due to their high elastic strain limit and toughness. Moreover, it is likely that some conventional material will be replaced with metallic glasses and composites in our daily life in the near future.

Basu et al. [25] clarified the possible atomic configuration in liquid alloys through detailed study of the structure and crystallization behavior of metallic glasses.

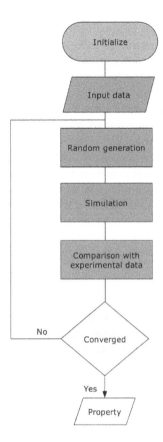

FIGURE 4.2   Flowchart of the procedure for calculation of thermophysical properties.

Due to the exploration of crystallization behavior, the synthesis of nanocomposite microstructures and their mechanical properties have been enhanced. Synthesis of bulk metallic glasses can be done through solidification and solid-state processing, in which water quenching is one of the oldest techniques in solidification. In this method, a vacuum-sealed quartz tube, in which alloy is melted, and then water is quenched, is used whereas modern techniques introduce arc melting, unidirectional zone melting, and injection molding for the generation of bulk metallic glasses. Metallic glass has considerably high plain strain fracture toughness, which indicates availability of plastic zone. Bulk metallic glass have innumerable properties, which have various applications in technical fields along with significant economic as well as environmental advantages. They also have various applications in aircraft frames, automobiles, armor penetrators, and medical implants due to their combination of high strength and toughness as well as magnetic properties.

Chatterjee et al. [26] has synthesized glass-metal nanocomposite through a metal organic route. They have also prepared glass-metal composites comprising nickel, cobalt, manganese, and ultrafine particles of iron with a silica-glass matrix through heat treatment of silicon tetra-ethoxide and metal organic compound. It was also found that the sol-gel technique is trending in the preparation of glass-metal nano-composites as well as other metallic particles.

During synthesis of bulk metallic glasses, the cooling rate has to be determined for the calculation of thermal conductivity and, during plastic deformation, local heating has to be estimated with narrow shear instabilities. It was also noticed that multi-element amorphous alloys have lower thermal conductivity than simple binary glasses. It was found that both phonons and electrons contribute to the thermal conductivity at room temperature. Experimentally, thermal conductivity has been deduced with the help of time dependence of temperature at two positions along the rod. To elucidate the thermal conductivity dependence on temperature, electrical conductivity was also measured on the same samples. It was concluded that at room temperature electronic contribution for thermal conductivity is proportional to temperature and phonon contribution is constant. Weidemann-Franz law was used to disintegrate the total thermal conductivity into electronic and phonon contributions, and it was observed that it provides a good measure of the electronic contribution to the thermal conductivity [27]

Figure 4.3 shows the schematic of the HRTEM structures, and Figure 4.4 shows the schematic of a metallic line deposited on a substrate.

Earlier rapid quenching of melt was utilized for making metallic glasses with a critical cooling rate along with limitations in the sample thickness. With the development of bulk metallic glasses, there is relaying in the interest of mechanical properties.

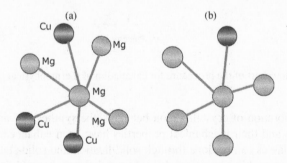

FIGURE 4.3   Schematic of HRTEM structures (a) with Mg and Cu (b) General structure.

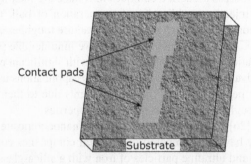

FIGURE 4.4   Schematic of a metallic line deposited on a substrate.

Metallic glasses have great potential as well as applications in tooling, springs, information storage, fashion items, micro electro-mechanical systems, etc. [28]

Michiaki Yamasaki et al. measured the thermal diffusivity as well as the conductivity of amorphous solid and $Zr_{55}Al_{10}Ni_5Cu_{30}$ bulk metallic glass with the help of the laser flash method experimentally. For finding thermophysical properties like thermal conductivity, the glass-forming mechanism should be cleared so that the critical cooling rate can be estimated. In studies of bulk metallic glasses, it is found that metallic glasses can be joined together through various welding techniques like friction, pulse current, laser beam, and electron beam welding. While evaluating the temperature variation of the thermal conductivity of $Zr_{55}Al_{10}Ni_5Cu_{30}$, it was found that, in the amorphous state, thermal conductivity as well as diffusivity were weakly temperature-dependent, with a small positive temperature coefficient. There were two discontinuous changes in thermal conductivity, one at the glass transition temperature and the other at the crystallization temperature, and these discontinuous changes were due to changes in specific heat capacity at glass transition temperature. Figure 4.5 shows the schematic of the experimental setup for obtaining the thermal conductivity. Thermal conductivity tends to increase and, when the temperature approaches 700 K, crystallization occurs. Moreover, Figure 4.6 shows the arrangement for the single impact of a MG particle onto a metallic substrate. Figure 4.7 shows interrelations for the formation of bioglass, biometallic alloys, and biometallic glass. It was found that the metallic glasses with a larger number of elements tend to show lower thermal conductivity and a higher phonon component [15].

Lv et al. [7] measured the thermal conductivity of Fe-based bulk metallic glasses and the internal friction behavior with different crystallization with a laser flicker thermal conductivity instrument and a multifunction, high-precision internal friction instrument. It was found that, with an increase in crystallization, thermal conductivity was gradually enhanced and, on the other side, the superplastic and crystallizing region was gradually weakened. The internal friction behavior of

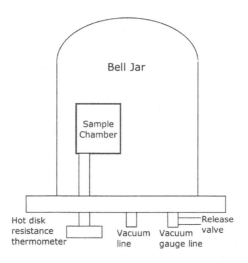

FIGURE 4.5 Schematic of the experimental setup for obtaining the thermal conductivity.

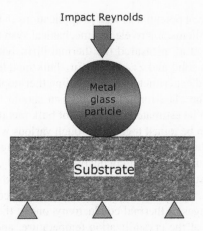

FIGURE 4.6   Arrangement for the single impact of a MG particle onto a metallic substrate.

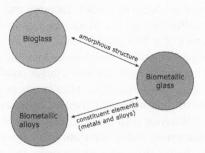

FIGURE 4.7   Interrelations for the formation of bioglass, biometallic alloys, and biometallic glass.

Fe-based bulk metallic glasses comprises three stages: high strength region, super-plastic region, and crystallization region, respectively. Moreover, through TEM investigation, it was found that medium-range order (MRO) was developed in the disordered atoms of bulk metallic glasses after partial crystallization whereas large number of dense nanocrystals were precipitated after complete crystallization, which may be one of the reasons for the increase in thermal conductivity. Compared to other bulk metallic glasses (BMGs) like Zr-based and Pd-based ones, Fe-based BMGs contain more comprehensive properties like high thermal stability and low thermal conductivity, which are exploited in building structure, refrigeration, and in petroleum engineering technology. Additionally, investigation of the thermal conductivity of Fe-based amorphous alloys is important for developing thermoelectric materials with enhanced properties. It was observed that, with an increase in annealing temperature, there was gradual enhancement in the thermal conductivity due to an increase in the extent of crystallization of amorphous alloys. The enhancement of thermal conductivity produced by MRO may be due to the difference in the average free path of phonons, and it was predicted that the higher the number of crystal domains, the higher the heat conduction efficiency. It is worth noting that the

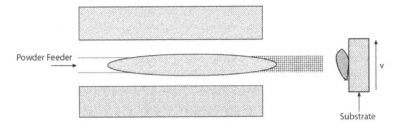

FIGURE 4.8   Schematic for the formation of substrate.

number of precipitated crystal domains was more closely related to the enhancement of thermal conductivity in BMGs. Figure 4.8 shows the schematic for the formation of substrate. Moreover, it was observed that, due to the heating process, there is an increase in the diffusion rate, which results in a change in structure of metastable atoms due to which free space gets vanished between the atoms, affecting the thermal conductivity of amorphous alloys.

Albe et al. [29] investigated the enhancing of plasticity of metallic glasses and nanocomposites using molecular dynamics computer simulations and studied the effect of microstructure in enhancing the strength of crystalline structure. Moreover, the effect of grain size and composition on nanoglass deformation behavior and nanocomposites was studied, and it was concluded that structural heterogeneities are shown by glass-glass interface, which initiates shear band formation and resists strain localization. Additionally, recent studies review those molecular dynamics simulations, which show the relation between the mechanical properties of metallic glasses with their atomic structure.

Nanocomposites can be synthesized by electrodeposition, in which ultrafine metal dispersions of copper, silver, and iron, respectively, have been grown in silica gel or glass matrix. Some nanocomposites were synthesized recently with the above technique, among which Ni-CeO$_2$ was a new kind nanocomposite to achieve a finer-size grain and higher microhardness as well as better wear resistance. Ni-CNT was also synthesized, and it was concluded that the mechanical properties and wear resistance of the composites were higher than that of particles of nanocrystalline nickel. Recently, glassy carbon electrodes were used for direct electrodeposition of gold nanostructure onto it. Furthermore, a strategy was used to synthesize an oxide glass with suitable treatment of heat and reduction, leading to the development of nanostructure containing a ferromagnetic core. This nanocomposite showed the characteristics of a multiferroic system. The glass composition which was used melted at 1473 K and was quenched through pouring it into an aluminium mold. Furthermore, the glass was reduced at 893 K into hydrogen, resulting in the formation of iron nanoparticles within the medium. Additionally, BaTiO$_3$ shells around Fe nuclei were formed after subsequent heat treatment at 759 K for four hours. Previously, for the manufacturing of ultraviolet resistance window glasses, glass-metal nanocomposites were used commercially by Pilkington Brothers, and many of the colored glass articles were made of composites of glass and metal nanoparticles. Due to the different resistivity value of glass ceramic metal nanocomposites, they can be utilized as electric circuits as well as humidity or gas sensors. Moreover, they have also an application in supercapacitors because of the ultra-high dielectric permittivity of metallic glass nanowire composite, which makes them a huge potential candidate for making energy storage devices. Additionally, nanoglass composites can be used as magnetic sensors in detecting magnetic fields. Furthermore, investigation of glass-based nanocomposites has discovered new areas for research like magnetoelectric phenomenon and in the field of solar cells generating voltage upon being exposed to solar radiation, which could be possible by proper heat or chemical treatment [30].

Tungsten has the highest melting point among other known metals, and its remarkable mechanical properties, high density, and high thermal conductivity make it, as well as its alloy-based composites, a candidate for high-temperature and structural applications. Moreover, to exploit metallic glasses for high-temperature applications, their crystallization temperature should be enhanced to resist any phase changes during operations, and thus tungsten is a refractory alloying element used to improve glass-forming ability and crystallization temperature. Moreover, it is practically difficult to form bulky glass systems containing refractory metals by casting, melting, and quick solidification techniques. However, studies has been carried out to investigate the fabrication of multicomponent $Ti_{40.6}Cu_{15.4}Ni_{8.5}Al_{5.5}W_{30}$ bulk metallic glasses with higher thermal stability using a mechanical alloying method, in which a solid-state reaction process is carried out using a low-energy ball-milling method, to form homogenous glassy powder, and then, by a spark plasma sintering approach, this powder is transformed into bulk metallic glasses. Furthermore, the powder obtained after 200 hours of milling process as a product consists of high thermal stability and is specified by 90 K for the supercooled liquid region and 827 K for the crystallization temperature. The metallic glasses obtained as a product consist of enhanced microhardness and high thermal stability [31].

Figure 4.9 shows the schematic representation of the experimental setup for the calculation of thermophysical properties of the specimen. Moreover, silver oxide glass nanocomposites are fair model systems to build a link between structure and transport properties in disordered materials. Furthermore, it was found that oxide glass nanocomposites manifest semiconducting behavior due to the presence of Mo ions in valence states in glassy matrix [32]. Glass nanocomposites require an essential condition for the formation from melt; the cooling should be rapid to prevent crystal nucleation and growth as well as more thermally stable crystalline phase. The requirement for glass nanocomposite to form is that the nucleation rate should be minimized below a value of $10^{-6} cm^{-1} s^{-1}$ [33]. Moreover, the splat quenching method is the quickest melt-quench technique, developed specially for metallic glasses by Duwez. However, when these glassy nanocomposites are passed through heat treatment higher than their glass transition temperature, it is found that the thermal conductivity of heat-treated samples is more than that of their glassy counterpart, and it is supposedly due to precipitation [34].

It has for quite some time been guessed that any metallic fluid can be vitrified into a shiny state given that the cooling rate is adequately high. Experimentally, however, vitrification of single-element metallic fluids is famously difficult. Genuine research of the development of monatomic metallic glass has been lacking. Here we report a

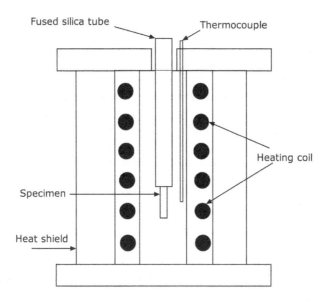

FIGURE 4.9 Schematic representation of the experimental setup for the calculation of thermophysical properties of the specimen.

trial way to deal with the vitrification of monatomic metallic fluids by accomplishing an uncommonly high fluid extinguishing rate of 10^{14} Ks^{-1}. Under such a high cooling rate, melts of unadulterated recalcitrant body-focused cubic (bcc) metals, for example, fluid tantalum and vanadium, are effectively vitrified to frame metallic glasses reasonable for property cross-examinations. The accessibility of monatomic metallic glasses, being the least difficult glass formers, offers exceptional potential outcomes for concentrated structure and property connections of glasses. Our procedure likewise indicates extraordinary authority over the reversible vitrification–crystallization forms, recommending its potential in small-scale electromechanical applications. The ultrahigh cooling rate, moving toward the most astounding fluid extinguishing rate achievable in the experiment, makes it conceivable to investigate the quick energy and basic conduct of supercooled metallic fluids inside the nanosecond to picosecond routines [35].

4.2   MODELING OF METALLIC GLASS AND THEIR NANOCOMPOSITES

A modeling or simulation workflow typically starts with the specification of an initial structure. For crystalline materials, this involves setting the correct lattice parameters, which ensures the short- and long-range orders which characterize those materials. Amorphous materials, on the other hand, lack long-range order, so a construction algorithm is required that generates the correct short-range correlation without introducing long-range order.

4.2.1   SIMULATION STRATEGY

The initial step before the prediction of thermal conductivity of metallic glass is the creation of its simulation model. Figure 4.10 shows the methodology adapted for performing MD simulation. Material Studio is one such commercial computational code utilized for molecular dynamics (MD) simulation and estimating its properties. MD simulations were performed, where Condensed-stage Optimized Molecular Potentials were utilized for Atomistic Simulation Studies (COMPASS). The COMPASS force field consists of terms for bonds (b), angles (θ), dihedrals (ϕ), out-of-plane angles (χ) as well as cross-terms and two nonbonded functions, a Coulombic function for electrostatic interactions and a 9–6 Lennard–Jones potential for van der Waals interactions.

$$E_{Total} = E_b + E_\theta + E_\phi + E_\chi + E_{b,b'} + E_{b,\theta} + E_{\theta,\theta'} + E_{\theta,\theta',\phi} + E_q + E_{vdW} \tag{4.1}$$

Initially, a single segment was built, preceded by the creation of a nanostructure (here, carbon nanotubes). A simulation cell is created, where the components of metallic glass and nanostructure were finally packed.

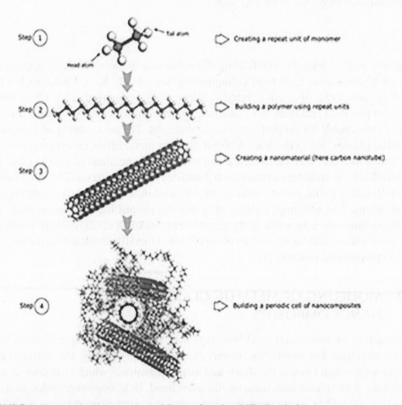

FIGURE 4.10   Methodology adapted for performing MD simulations.

4.3   BFGS GEOMETRY OPTIMIZATION

The ability to perform cell optimization, along with fixed external stress optimization, is the main advantage of the BFGS minimizer. The use of finite basis set correction is recommended when cell optimization is required. The calculation of the first term is not computationally expensive (the first iteration involves 10%–30% of the self-consistent electronic minimization) relative to the advantages provided by it. However, when the energy cut-off is low and finite basis set correction is not utilized, cell optimization may be problematic. In such circumstances, the optimization process may stop due to a nonzero value of stress being generated while the energy converges. However, the minimizer attempts for energy minimization rather than stress zero-point since the former is desired under many circumstances.

4.3.1   CONSTRAINTS

During the geometry optimization process, constraints are employed in order to fix atom positions. In the molecular dynamics approach, this implies the fixation of the fractional coordinates of the constituent atoms. Moreover, it is also possible to fix the lattice parameters. Such constraint is useful when dealing with phase stability studies. It is also possible to implement linear constraint on atomic coordinates. Such constraint is implemented using transformation matrix, which changes the Cartesian coordinates of all coordinates to subspace of unconstrained coordinates. For example, for slab calculations of surface processes, such provision is useful.

4.3.2   NON-LINEAR CONSTRAINTS

Constraints such as angles, inter-atomic distances (bonds), and torsions are referred to as non-linear constraints. By the utilization of a delocalized internals optimizer, such constraints can be imposed.

4.3.3   ESTIMATED COMPRESSIBILITY

By making a reasonable initial estimate of the bulk modulus for the materials, the cell optimization convergence can be made. The initial changes in lattice vectors are calculated on the basis of this value. The steps become smaller for the minimizer when the estimated bulk modulus is too high; therefore, it is possible for reduction in the number of geometry optimization steps by reducing this value. However, oscillating behavior will occur if the bulk modulus for a hard material is too small. Thus, for reduction in the number of time steps required for calculations, prior knowledge of the hardness of the material is important.

4.3.4   TPSD GEOMETRY OPTIMIZATION

Assuming that the potential is quadratic, the BFGS algorithm determines the optimal step length per iteration. However, a more robust two-point steepest descent (TPSD) Barzilai-Borwein algorithm allows a wider range of problems to be solved. When a cell is optimized with user-supplied constraints, the two-point steepest descent (TPSD) algorithm allows a wider range of problems of be solved. For example, in the optimization of solid-solid interfaces strained to a substrate, all the cells angles and lateral cell parameters are fixed and the only relaxation allowed is in the direction normal to the interface. However, only the position and gradient of the system at the current and previous iteration is required by the TPSD algorithm, which implies a relatively small amount of memory will be consumed. A step size is calculated based on the difference of each value between the iterations, for usage in the direction of the most negative gradient from the current iteration. A reliable convergence is found in the TPSD optimization on a number of occasions where the BFGS takes a long time to converge.

4.3.5 Damped Molecular Dynamics

An alternative method for geometry minimization is presented by damped molecular dynamics, which involves only internal coordinates (cell parameters) to be fixed. For dealing with the ground state, a critical damping regime is utilized for dealing with the ground state. For implementation of this regime, either one damping coefficient is implemented for all degrees of freedom or distinct coefficients are used for different degrees of freedom. Both the modes of damped molecular dynamics can be performed with a larger time step than undamped molecular dynamics simulations. The time step can be increased further when the system gets close to equilibrium since the fast modes freeze out before the slower ones. Higher efficiency of the algorithm is attained due to the automatic adjustment of the time step.

4.3.5.1 The Optimization Process

The optimization of a structure is a two-step process:

* Energy evaluation

For every given conformation, the energy expression must be suitably defined. Combined with force field terms, the coordinates of the structure are obtained by creating an energy expression (or target function). For a particular structure, the potential energy surface is defined by the energy expression. The potential energy of a system of atoms can be expressed as a sum of cross terms, valence (or bond), and nonbonded interactions.

$$E_{total} = E_{valence} + E_{crossterm} + E_{nonbond} \tag{4.2}$$

* Conformation adjustment

To reduce the value of energy expression, conformation is adjusted. Depending on the algorithm, either one adjustment or thousands of iterations may be required. The time required to evaluate the energy expression and the number of structural adjustments (i.e., iterations) determines the efficiency of optimization and time required for convergence to the minimum.

4.3.6 GEOMETRY OPTIMIZATION

An initial step before any dynamic analysis is to attain minimum energy of the obtained molecular structure for stability purposes. The high-energy configuration of the obtained geometry can lead to erroneous results unless the energy minimization task is performed. Under this scheme, all possible positions of the molecules are analyzed and the configuration corresponding to the overall lowest energy state is selected for further analysis.

An essential part for the modeling of molecular simulation is geometry optimization. For such a process, linear and exponential methodologies are available, which implies the diversity of scaling methods. Two general requirements are present for obtaining optimized geometry, reliability and good scaling with size. Over the past three decades, rapid evolvement of geometry optimization techniques has taken place. One of the major developments was the inclusion of analytic gradients of the potential energy and the methodology based on them, such as the quasi-Newton methods. For the acceleration of the optimization process, the Hessian update techniques have a provision for collecting information about the potential energy surface (PES). The existing techniques, such as rational function optimization (RFO), quadratic line search (QLS), trust radius update, and trust radius model (TRM), made such improvement achievable. However, the geometry optimization was originally developed for Z-matrix internal coordinates or in Cartesian coordinates. It was only in the last decade that the redundant internal coordinates took precedence. Improvements using other techniques, such as delocalized or natural internal coordinates or geometry optimization using direct inversion in the iterative subspace (GDIIS), were also considered. However, for quantum chemical applications, it is now commonly agreed that an efficient optimization method could be achieved by utilizing GDIIS or line search, Hessian update, and RFO in the framework of redundant internal coordinates. Moreover, general use of the optimization techniques is prevented by the quadratic ($O(N^2)$) due to scaling of their memory and the number of variables to optimize due to scaling of their computational demand in $O(N^3)$. Quantum scaling methods, like semi-empirical, density functional theory, and Hartee-Fock are current advances in scaling, which necessitate the development of new optimization techniques for matching the required scaling with size which does not compromise the traditional efficiency and reliability. A reduction to asymptotic quadratic, $O(N^2)$, can be achieved using overall computational bottleneck, where scaling using updated inverse techniques can be used for solving equations, subjected to both TRM optimization technique and coordinate transformations and the RFO.

In simulation software, such as Materials Studio, the CASTEP module is utilized for achieving geometry optimization. Two schemes are supported by CASTEP for geometry optimization, BFGS and damped molecular dynamics. Hessian is used by the BFGS scheme initially, which is updated during optimization recursively. For optimizing both atomic parameters and lattice parameters, Hessian is implemented in the mixed space of internal cell and degrees of freedom.

4.3.7 THERMAL CONDUCTIVITY

Heat transport in materials is an important property that is quantified through thermal conductivity. Because this depends on phonon-phonon scattering processes, the usual methods to calculate this property typically involve molecular dynamics simulations. However, a number of approximate schemes have been proposed that work within the framework of lattice dynamics. Here, the phonons are calculated for a supercell and, by broadening the phonon density of states using a Lorentzian, the thermal conductivity, κ, can be computed according to the following expression:

$$\kappa = \frac{1}{V} \sum_{i=1}^{modes} C_i(T) D_i$$

(4.3)

where V is the cell density volume, $C_i(T)$ is the heat capacity of mode i at temperature T, and D_i is the model diffusivity.

4.4 PREDICTION OF THERMAL CONDUCTIVITY

With the mechanical instrument and devices toward the nanoscale, productive heat extraction is of essential significance for their proper functioning. While the heat transport phenomenon dielectric has been obtained, many issues regarding nanocomposites are still uncertain. Moreover, the contribution of individual deformities leads to difficulty in interpretation of the obtained experimental results. Hence, molecular dynamics (MD) simulators provide an appealing approach for prediction of properties of materials [36]. The relation between thermal conductivity and heat current is given by

$$J_\mu = -\sum_v \kappa_{\mu v} \frac{\partial T}{\partial x_v}$$

(4.4)

where J_μ is a thermal current, $\kappa_{\mu v}$ is the thermal conductivity tensor, and $\partial T/\partial x_v$ is the temperature gradient. The value of κ is obtained by measuring the gradient in temperature under the application of a heat current.

4.4.1 MODELS FOR PREDICTING THERMAL CONDUCTIVITY

Many theoretical and semi-empirical models are available for obtaining thermal conductivity of composite materials. A few of them have been mentioned in the preceding sections.

4.4.2 PARALLEL AND SERIES MODELS

In this methodology, both the models assume independent contribution to the overall conductance and thermal resistance. In the case of laminated stacked composite, a perfect interface between the phases in contact is assumed. The thermal conductivity is given by

$$k = \frac{\left(k_m \times k_f\right)}{\left(k_f \times \left(1 - v_f\right) + k_m \times v_f\right)} \tag{4.5}$$

where k_f represents the thermal conductivity of nanofillers, k_m denotes the thermal conductivity of nanocomposite matrix, and v_f represents the nanofillers volume fraction.

4.4.2.1 Maxwell–Garnett Model

The issue of determining the transport properties of multiphase materials goes back to Maxwell. In view of the progression of potential and electric flow at the interface, and on the assumption that the interactions among the strands of fiber can be neglected, which implies that the filaments are situated sufficiently far from one another, Maxwell analytically determined an equation for the effective specific resistance (K) of "a compound medium consisting of a substance of specific resistance K_2, in which are disseminated small spheres of specific resistance K_1, the ratio of the volume of all the small spheres to that of the whole, being p." When changed to the thermal conductivity (k), the model gives:

$$k = k_m + \left[3 \times v_f \times \frac{\left(k_f - k_m\right)}{2 \times k_m + k_f - v_f \times \left(k_f - k_m\right)}\right] \tag{4.6}$$

4.4.2.2 Lewis–Nielsen Model

In this model, the flux distribution becomes analogous to stress and conductivity becomes analogous to stiffness. The thermal conductivity is modeled as:

$$k = k_m \times \left[\frac{1 + A \times B \times v_f}{1 - B \times \psi \times v_f}\right] \tag{4.7}$$

where

$$A = k_E - 1 \tag{4.8}$$

$$B = \frac{\dfrac{k_f}{k_m} - 1}{\dfrac{k_f}{k_m} + A} \tag{4.9}$$

$$\psi = 1 + 1.775 \times v_f \tag{4.10}$$

k_f represents the thermal conductivity of nanofillers, k_m depicts the thermal conductivity of nanocomposite matrix, and v_f represents the volume fraction.

4.4.2.3 Hamilton–Crosser Model

The Hamilton and Crosser model [30] is an expansion of Maxwell's hypothesis representing the nonsphericity of fillers using a shape factor (n), characterized as $n = 3/\psi$, with ψ being the molecule sphericity. The sphericity $(\psi = Ae/An)$ of a molecule is characterized as the proportion of the surface region (Ae) of the proportionate circle having a similar volume to the real surface region (An) of the noncircular fiber. The viable thermal conductivity of the Hamilton and Crosser model is given by:

$$k = k_m \left[\frac{k_f + 5k_m - 5v_f \times \left(k_m - k_f\right)}{k_f + 5k_m + v_f \times \left(k_m - k_f\right)} \right] \qquad (4.11)$$

where k_m = the thermal conductivity of polycarbonate matrix = 0.15 W/m/K, k_f = the thermal conductivity of armchair MWCNT = 30 W/m/K, v_f = the MWCNT volume fraction.

4.4.2.4   Deng Model

For nanofiller composites with low loadings of normal length L and measurement d, the expression for thermal conductivity is

$$\frac{k_e}{k_m} = 1 + \frac{\dfrac{v_f}{3}}{\dfrac{k_m}{k_f} + H} \qquad (4.12)$$

where k_e represents the equivalent thermal conductivity of the composite, k_m represents the thermal conductivity of nanocomposite matrix, k_f depicts the thermal conductivity of nanofillers, and v_f represents the nanofillers volume fraction,

$$H = \frac{1}{p^2 - 1} \left[\frac{p}{\sqrt{p^2 - 1}} \ln\left(p + \sqrt{p^2 - 1}\right) - 1 \right] \qquad (4.13)$$

$$p = \frac{l}{d} \qquad (4.14)$$

4.5   CURRENT TRENDS AND APPLICATIONS

Bulk metallic glasses (BMG) possess interesting properties, which can be employed for various industries and transportation purposes. High tension and compression strength is exhibited in them, along with hardness, elastic elongation limit, and corrosion resistance. Advancement in the field of miniature applications has taken place. Bulk metallic glass, when employed in the range of length scale ~1–5 mm benefits the most. These fields of applications include NEMS (nanoelectromechanical systems), micro parts, micro machines, MEMS (microelectromechanical systems), and surgical tools.

Moreover, BMG alloys are being produced in the form of composites, such as sheets, rods, pipes, and plates, for structural applications. The main attributes of BMG composites are low Young's modulus, high fracture strength, and high yielding strength, which are utilized for production of parts with intricate shapes, such as coiled springs and gears. Moreover, the corrosive resistance of BMG over crystalline structure plays a significant role in chemical industries. Furthermore, BMG composites have been considered as a suitable material for the manufacturing of fuel cell separators. Recently, progress has been made in the development of proton-exchange membrane fuel cells (PEMFCs) by employing the viscous deformability and corrosion resistance properties. Moreover, in the field of biomedical applications, implants which biocorrode are of particular interest. Utilizing BMG composites in surgical equipment and devices decreases the long-term negative effects of implants. In the sector of electrical applications, the magnetic properties of Fe-based BMGs have been employed in power distribution transformers.

Bulk metallic glasses (BMGs) have already been commercialized to be used in different applications, and still it is being explored due to its peculiar characteristics. High compression and tensile strength, microhardness, electrical resistivity, good soft magnetic properties, excellent corrosion resistance, and high biocide effect, etc. make bulk metallic glass (BMG) and its composites the center of attraction for all scientists and researchers. Day by day, researchers are getting involved in experiments with this novel material to study the feasibility of using this material for some advanced applications. In most of the cases, it is being found that using bulk metallic glasses is more advantageous than using any other kind of conventional materials. Some advanced trends of using this material are summarized.

Choi-Yim et al. [37] made three different types of compositions of bulk metallic glass alloys of the compositions of $Cu_{47}Ti_{34}Zr_{11}Ni_8$ (V101), $Zr_{52.5}Ti_5Al_{10}Cu_{17.9}Ni_{14.6}$ (V105), and $Zr_{57}Nb_5Al_{10}Cu_{15.4}Ni_{12.6}$ (V106) with the solid secondary phase materials, such as SiC, WC or TiC, used as ceramics and metals used as W and Ta. The volume concentration used varied from 5% to 30% with the size of the particles in between 20 μm and 80 μm. The composite ingots then melted in the temperature range of 850°C–1100°C in a quartz tube under a vacuum using an induction heating coil. Then it was injected into an argon-based copper mold at 1 atmospheric pressure. It was observed that the secondary phase did not affected the glass-forming abilities of the composite, and it is uniformly distributed over the matrix. Thus, it can be concluded that this method of producing matrix is simple and effective.

It was found by Wang et al. [38] that a novel biodegradable material (CaMgZn bulk metallic glass) could be promising for potential skeletal applications. They experimented on mice and no mice died, and there is no report of inflammation observed. It could be seen that, the material is biodegradable in Hank's Solution, and it has also been noted that it is degrading as fast as 3 hours. It has also been seen that, after four weeks of implantation, a little amount of residue of composite was left in the confined area of the bone marrow cavity of the mouse. It has also been found that, in a wide range of concentrations, the material showed a good biocompatibility with the L929, VSMC, and ECV304 cells, and it is promoting the viability of MG63 cells and producing alkaline phosphate (ALP) in the concentration range of 10%–30%, which is a benefit for new bone formation; however, at higher concentrations, it can change the MG63 cell morphology and cause apoptosis and, at lower concentrations, MG63 cells incubated with the material extracts and have a well-stretched F actin distribution and clear nuclei. As the Young's modulus is so close to the human bone, it could be a novel material to be used in orthopedics as screws or plates, etc. Calin et al. [39] investigated titanium-based metallic glasses for biomedical implant applications. They observed better corrosion resistance, high fracture strength and wear resistance, increased elastic strain range, lower Young's modulus, and biocompatible characteristics, leading to the use of titanium-based metallic glass composites for several biomedical applications. They designed two compositions, i.e., ternary $Ti_{75}Zr_{10}Si_{15}$ and quaternary $Ti_{60}Nb_{15}Zr_{10}Si_{15}$, and evaluated glass-forming ability, thermal stability, and mechanical and corrosion behavior. It was seen that Ti–(Nb)–Zr–Si metallic glasses and nanocomposites show excellent performance, including high hardness and strength, high specific strength, and excellent corrosion resistance, with passive current density one order of magnitude lower than that of cp-Ti. It could also be seen that $Ti_{60}Nb_{15}Zr_{10}Si_{15}$ fully amorphous ribbon provides a more enhanced protective effect. Thus, it could be concluded that this nanocomposite shows better qualities than any other conventional materials.

Murr et al. [40] tested thin sheets and small wires of $Fe_{80}B_{20}$, and thin sheets of $Fe_{38}Ni_4Mo_4B_{18}$ glassy metals were explosively shock loaded together with thin SS (stainless steel) 304 sheets at peak pressure pulses of 15, 25, and 35 GPa at a shock pulse duration of 2 μs. It was observed that no significant changes in residual hardness happened even for the highest peak loads. There were no changes in their structure as well as properties, leading this novel material to be used in a wide range of applications (defense, space technology, etc.) where there is a need for shock-absorbing materials. Very short range order (microcrystallites) or regular geometrical arrays were observed, which were composed of random structure.

Gloriant [41] studied microhardness and abrasive wear resistance of lanthanum-based, zirconium-based, and palladium-based metallic glasses on two Al-based amorphous ribbons. It was observed that hardness and wear resistance was enhanced significantly, leading to their use in high wear applications. Kumar et al. [42] explained how bulk metallic glass (BMG) composites can be used in manufacturing MEMS, NEMS, various micro-sized parts, micromachines, bioapplications, and several other precision tools. Kobayashi et al. [43] introduced a new method of production of bulk metallic glass composites by using smart thermal spray, which is more cost-effective than other conventional methods. It was also observed that the hardness of the metallic glass coating was increased with the increase of plasma current. For these qualities, this could be used as structural components very effectively. Siegrist et al. [44] investigated $Zr_{52.5}Cu_{17.9}Al_{10}Ni_{14.6}Ti_5$ BMG reinforced with 25–44 μm and, in another case, 44–75 μm graphite particles. Most of the BMG composites show very high yield strength but very low plastic deformation before fracture. To improve plastic deformation characteristics, this kind of composite was introduced. It can be seen that plastic strain improved up to 18.5% in the case of this BMG composite without sacrificing the yield strength (1.85 GPa). These results have the highest combination of yield strength and compressive plastic strain so far reported in foreign particle-reinforced bulk metallic glass composites. Hence, it could be a more excellent structural material than other conventional materials.

Li et al. [45] numerically analyzed the ballistic characteristics of tungsten particle/metallic glass composites and tungsten fiber/metallic glass composites over conventional tungsten heavy alloy and depleted uranium alloy. It can be seen from their results that the composites, show enhanced penetrating characteristics. For both the composites, the penetrating power is somewhat the same, and both show self-sharpening characteristics and thus increased penetration. However, in the case of tungsten particle/metallic glass composites, the effect of erosion is high and thus penetration is a little lower than that of the tungsten fiber/metallic glass composites. It can be observed from their work that, in most of the factors of ballistic characteristics, like impact velocity, target strength, and the initial nose shape, the composite shows influenced and enhanced performance, leading to its use in the manufacturing of missiles bullets, etc.

González et al. [46] experimentally analyzed the antimicrobial and wear performance of $Cu_{50}+_x(Zr_{44}Al_6)_{50-x}$ with $x = 0, 3, 6$ composites. It can be clearly observed that these composites have a good number of antimicrobial effects. There was a significant reduction in colony-forming units for *Escherichia coli* (gram-negative) and *Bacillus subtilis* (gram-positive) after 60 minutes of contact time for the $Cu_{56}Zr_{38.7}Al_{5.3}$ alloy and, at 250 minutes, the complete elimination of bacteria can be found in their work. It assures the use of this type of composite in the health sector as well as in any sector as health is a primary safety concern for all aspects. Similarly, Huang et al. [47] investigated Al-Ni-Y-based thin-film metallic glass composites for their optical reflect ability and antibacterial and antifungal characteristics. It was demonstrated that the composite shows very high reflective nature (83%–89%) under the ultraviolet region, and it shows around 99.99% antibacterial and level 0 antifungal capability. Also, its very high hardness (~6 GPa) and extremely low surface roughness (1–3 nm) makes the material useful in high-humidity conditions with long-term and efficient use. Steyer et al. [48] demonstrated enhanced physiochemical properties of $Zr_{39}Cu_{39}Ag_{22}$ thin-film metallic glass deposed by magnetron PVD sputtering. It was found that this composite shows very good biocide effect against gram-positive. According to JIS Z2801:2000 standard, it also shows high Young's modulus (95 GPa), high hardness (5 GPa), high temperature stability up to 250°C, and very good anticorrosive properties. They have proposed to use this kind of composite to be used in several biomedical applications.

Kozachkov et al. [49] in their precious work described how bulk metallic glass composites can be used as spacecraft shielding as the number of micrometeoroid and orbital debris (MMOD) collisions with spacecraft and satellites has been increasing with the use of lower earth orbit. Singh et al. [50] studied the joining of three $ZrBr_2$-based ultrahigh-temperature ceramic composites using metallic glass interlayers with the use of two high-temperature amorphous Ni-based braze alloys containing boron whose liquidus temperatures is around 1050°C for using in excess temperatures of 1600°C. Three specimens consisting of ZrB_2–SiC–C (ZSC), ZrB_2–SCS_9–SiC (ZSS), and ZrB_2–SiC (ZS) are joined with brazes (Met-glass MBF-20 and MBF-30). It was found that ZS composites exhibited high hardness (around 2100 KHN) after the joining process at very high temperature. They demonstrated that this composite can be used reliably at higher temperatures in various high-temperature applications.

Hsieh et al. [51] studied reflectivity of ultraviolate (UV) range radiation over polydimethylsiloxane (PDMS) lens coated with aluminium-based thin-film metallic glass (TFMG). It was observed that this novel material has significant potential to replace the conventional mercury (Hg) lamp exposure system, which is the finest environmentally friendly material. It can be seen that using aluminium-based thin-film metallic glass coating over the lens as reflector is enhancing the reflectance at ultraviolet wave bands. It can also be observed that the leakage of radiation at the large angle portions of lens can be reduced and can be concentrated to the target plane through this new material. The average irradiance can be enhanced by 6.5%–6.7%. The uniformity of the radiation was also improved; the output power is also enhanced; and the dimensional deviation was reduced around 3%. This kind of novel material can be used in the exposure system in the micro-electronic process, such as the printed circuit industry. Panet et al. [52] designed a surgical shadowless lamp with a multifocal, ellipsoidal curve reflector, and various recipes of Ag–Cu–Al thin-film metallic glass (Ag-based TFMG) were determined and coated on the curve as reflector materials, which possesses great mechanical and anti-microbial characteristics and is able to absorb the infrared range of radiation. This kind of LED-based reflection type shadowless surgical lamp with multifocal ellipsoidal curve layers could replace traditional projection-type surgical shadowless lamps, which would be cost-effective with the reduction of the number of light lamps. Huang et al. [53] explored the reflection characteristics of different types of Al-Ni-Y-X(X=Cu, Ta, Zr)-based thin-film metallic glasses over broad optical bands. It was found that the reflectance of the composite varies within 80%–91%, which is adequate. The surface roughness and hardness are also enhanced along with the advantage of an antimicrobial effect in comparison with that of the pure aluminum-based reflector. The reflectance of the BMG composite is nearly constant over the ultraviolet and infrared range of optical wavelength, with reflectance varying from 80% to 91%. It can be found that over the wide range of wavelength (from 200 to 1000 nm) the composite shows highly uniform reflection at around 70% with less spectroscopic distortion.

Tian et al. [54] explored the electrochemical behavior of bulk metallic glass composites for their use as bipolar plates for proton exchange membrane fuel cell (PEMFC) stacks. They considered $Zr_{55}Cu_{30}Al_{10}Ni_5$ bulk metallic glass in the solution of 0.05 M H_2SO_4 + 2 ppm F^- and they studied the effect of potentiostatic polarization on the interface contact resistance between the bulk metallic glass and carbon paper. It was observed that the bulk metallic glass undergoes excellent passivation and the passive film is very stable under the electrochemical operation between anode and cathode at around 70°C. It was also found that corrosion takes place at the interface of the $Zr_{55}Cu_{30}Al_{10}Ni_5$ during potentiostatic polarization at the specified potential, and no pitting was found at the interface. The interface contact resistance and compaction force at the $Zr_{55}Cu_{30}Al_{10}Ni_5$ was found to be 8 mΩ-cm^2, and 150 N/cm^2, respectively, which is much lower than the other conventional materials. Thus, it can replace the conventional bipolar plate material for PEMFC.

Wang et al. [55] investigated physical and mechanical properties of zirconium-based bulk metallic glass reinforced with carbon nanotube (CNT). Experimental results show that the compressive strength is high with high fracture strength. It was found that the density of the BMG composite reduces with the increasing volume concentration of CNT, but longitudinal and transverse acoustic velocities, Young's modulus, bulk modulus, shear modulus, and hardness increase with the volume concentration of the CNT. It was noted that for 4% volume concentration of CNT reinforcement, relative changes in longitudinal acoustic velocity, transverse acoustic velocity, Young's modulus, bulk modulus, and rigidity modulus were up to 11.11%, 6.39%, 11.21%, 23.91%, and 10.86%, respectively. These results indicate the strengthening ability of the composites with the increase of volume concentration of the CNT. It was also noted that the composites have excellent wave absorption ability with very high ultrasonic attenuation as the coefficient of attenuation was very sensitive to the volume concentration of the CNT and increased non-linearly with the volume concentration of the CNT. The ultrasonic attenuation and wave absorption properties built due to the scattering of crystalline ZrC phase mixed with CNT dispersed in the glass matrix, and the interface built between the glassy phase and CNT. These results indicate that this material can be efficiently used for shielding of noise and can play a major role to reduce the environmental noise pollution.

Tregilgas [56] made a new kind of digital light processor. In this precious work, amorphous titanium-aluminide-based material was used and it was seen that it allows the rotation of each micromirror, and the life cycle was increased significantly. It was announced the winner of the 2004 ASM Engineering Materials Achievement Award. This technology has allowed digital cinema to become a reality. Dandliker et al. [57] demonstrated how to use bulk metallic glass composites as a penetrator efficiently. The amorphous composite comprises 41.25% atomic percent zirconium, 41.25% titanium, 13.75% copper, 12.5% nickel, and 22.5% beryllium. An excellent property exhibited by this composite is its use as a kinetic energy penetrator. Gorsse et al. [58] investigated the magnetocaloric effect and refrigeration capacity of $Gd_{60}Al_{10}Mn_{30}$ nanocomposite. It was observed that it exhibits excellent refrigeration capacity with the soft magnetic characteristics, which allows it to be used in a wide range of temperatures. The ability to utilize this material as a magnetic refrigerant at 150 K is making the other materials a center of attraction.

Zhu et al. [59] studied the thermoelectric properties of $Ge_{20}Te_{80}$ nanocomposites. They heated the material and quenched it in water; for another sample, they quenched with liquid nitrogen, and other samples were produced by annealing at different temperatures. It was found from XRD patterns that fully crystalline material was formed for the water-quenched sample and a fully amorphous structure was found for the LN-quenched sample. Nanocomposites formed after annealing. It was revealed after observing various thermoelectric parameters that nanocomposites annealed at 443 K for 2 hours exhibit very good thermoelectric properties, which is cost-effective and efficient in today's scenario.

Inoue et al. [60] demonstrated how bulk metallic glasses (BMGs) can be used in different applications. They demonstrated that using $Fe_{73}Ga_4P_{11}C_5B_4Si_3$ BMG plates, 30 mm long, 20 mm wide, and 1 mm thick (prepared by squeeze casting) is very useful for the magnetic linear actuator applications. It was observed that it is very precise and efficient in the application range of 20–45 Hz frequency and it is showing good acceleration and deceleration characteristics, which is indicative by exhibiting large Lorenz force. It also shows very good magnetic shielding characteristics to be used in laptops and PCs. They also demonstrated how it can be used in PEMFCs (proton exchange membrane fuel cells) as it reduces the weight and cost and interfacial contact resistance, which can be seen in the work of Tian et al. [54]. They also mentioned using this material as structural material, MEMS (microelectromechanical system), and NEMS (nanoelectromechanical system), and optical material, which was validated by other researchers [28,39,44,47,51–53,56]. In addition to these, Inoue et al. [60] demonstrated that BMG materials as diaphragms of various strain gauge–based pressure sensors, as pipes for Coriolis mass flow meters, and as various parts for aircrafts and automobile valve springs.

It can be observed that bulk metallic glasses are replacing conventional materials with lots of beneficial properties and characteristics suitable for several applications, and still this material is being explored with varying the recipes and other properties of it can be used in some specific applications with better gainful characteristics and properties. Thus, BMGs are the only solution for various materialistic challenges for the future.

REFERENCES

1. Jun, W.K., Willens, R.H., Duwez, P.O.L. (1960), "Non-crystalline structure in solidified gold–silicon alloys," *Nature*, Vol. 187, p. 869.
2. Tahaoğlu, D. and Durandurdu, M. (2018), "Solute aggregation in $Ca_{72}Zn_{28}$ metallic glass," *Journal of Non-Crystalline Solids*, Vol. 500, pp. 410–416.
3. Qiu, Y.H., Xu, C., Fu, E.G., Wang, P.P., Du, J.L., Hu, Z.Y., Shao, L. (2018), "Mechanisms for the free volume tuning the mechanical properties of metallic glass through ion irradiation," *Intermetallics*, Vol. 101, pp. 173–178.
4. Lu, Y., Huang, Y., Wu, J., Lu, X., Qin, Z., Daisenberger, D., Chiu, Y.L. (2018), "Graded structure of laser direct manufacturing bulk metallic glass," *Intermetallics*, Vol. 103, pp. 67–71.
5. Zhang, X., Mei, X., Wang, Y., Wang, Y., Sun, J. (2018), "The study of irradiation damage induced by proton in metallic glass Ni62Ta38 and metal W," *Nuclear Instruments and Methods in Physics Research Section B: Beam Interactions with Materials and Atoms*, Vol. 436, pp. 1–8.
6. Deb Nath, S.K. (2018), "Thermal conductivity and mechanical properties of Zr_xCu_{90-x} Al_{10} under tension using molecular dynamics simulations," *International Journal of Mechanical Sciences*, Vol. 144, pp. 836–841.
7. Lv, Y. and Chen, Q. (2018), "Internal friction behaviour and thermal conductivity of Fe-based bulk metallic glasses with different crystallization," *ThermochimicaActa*, Vol. 666, pp. 36–40.
8. Presz, W. and Kulik, T. (2019), "Ultrasonic vibrations as an impulse for glass transition in microforming of bulk metallic glass," *Archives of Civil and Mechanical Engineering*, Vol. 19, pp. 100–113.
9. Mistarihi, Q.M., Hwang, J.T., Ryu, H.J. (2018), "The modelling and simulation of the thermal conductivity of irradiated U-Mo dispersion fuel: Estimation of the thermal conductivity of the interaction layer," *Journal of Nuclear Materials*, Vol. 510, pp. 199–209.
10. Kbirou, M., Trady, S., Hasnaoui, A., Mazroui, M. (2018), "Short and medium-range orders in Co_3Al metallic glass," *Chemical Physics*, Vol. 513, pp. 58–66.
11. Lu, Y., Huang, Y., Wu, J. (2018), "Laser additive manufacturing of structural-graded bulk metallic glass," *Journal of Alloys and Compounds*, Vol. 766, pp. 506–510.
12. Babilas, R., Nowosielski, R., Pawlyta, M., Fitch, A., Burian, A. (2015), "Microstructural characterization of Mg-based bulk metallic glass and nanocomposite," *Materials Characterization*, Vol. 102, pp. 156–164.
13. Garnier, B. and Boudenne, A. (2016), "Use of hollow metallic particles for the thermal conductivity enhancement and lightening of filled polymer," *Polymer Degradation and Stability*, Vol. 127, pp. 113–118.
14. Boussatour, G., Cresson, P.Y., Genestie, B., Joly, N., Brun, J.F., Lasri, T. (2018), "Measurement of the thermal conductivity of flexible biosourced polymers using the 3-omega method," *Polymer Testing*, Vol. 70, pp. 503–510.
15. Yamasaki, M., Kagao, S., Kawamura, Y. (2005), "Thermal diffusivity and conductivity of $Zr_{55}Al_{10}Ni_5Cu_{30}$ bulk metallic glass," *ScriptaMaterialia*, Vol. 53, pp. 63–67.

16. Diallo, M.S., Srinivasan, S., Chang, B., Ghosh, S., Balasubramanian, G. (2016), "Effect of metallic nanoparticle fillers on the thermal conductivity of diatomaceous earth," *Physics Letters A*, Vol. 380, pp. 3645–3649.

17. Zadorozhnyy, V.Y., Gorshenkov, M.V., Churyukanova, M.N., Zadorozhnyy, M.Y., Stepashkin, A.A., Moskovskikh, D.O., Kaloshkin, S.D. (2017), "Investigation of structure and thermal properties in composite materials based on metallic glasses with small addition of polytetrafluoroethylene," *Journal of Alloys and Compounds*, Vol. 707, pp. 264–268.

18. Henao, J., Bolelli, G., Concustell, A., Lusvarghi, L., Dosta, S., Cano, I.G., Guilemany, J.M. (2018), "Deposition behaviour of cold-sprayed metallic glass particles onto different substrates," *Surface and Coatings Technology*, Vol. 349, pp. 13–23.

19. Chen, X., Xiao, J., Zhu, Y., Tian, R., Shu, X., Xu, J. (2017), "Micro-machinability of bulk metallic glass in ultra-precision cutting," *Materials & Design*, Vol. 136, pp. 1–12.

20. Tlili, A., Pailhès, S., Debord, R., Ruta, B., Gravier, S., Blandin, J.J., Giordano, V.M. (2017), "Thermal transport properties in amorphous/nanocrystalline metallic composites: A microscopic insight," *ActaMaterialia*, Vol. 136, pp. 425–435.

21. Li, N., Zhang, J., Xing, W., Ouyang, D., Liu, L. (2018), "3D printing of Fe-based bulk metallic glass composites with combined high strength and fracture toughness," *Materials & Design*, Vol. 143, pp. 285–296.

22. Zhang, X., Mei, X., Zhang, Q., Li, X., Qiang, J., Wang, Y. (2017), "Damage induced by helium ion irradiation in Fe-based metallic glass," *Journal of Nuclear Materials*, Vol. 490, pp. 216–225.

23. Choy, C.L., Leung, W.P., Ng, Y.K. (1989), "Thermal conductivity of metallic glasses," *Journal of Applied Physics*, Vol. 66, pp. 5335–5339.

24. Li, H.X., Lu, Z.C., Wang, S.L., Wu, Y., Lu, Z.P. (2019), "Fe-based bulk metallic glasses: Glass formation, fabrication, properties and applications," *Progress in Materials Science*, Vol. 103, pp. 235–318.

25. Basu, J., and Ranganathan, S. (2003), "Bulk metallic glasses: A new class of engineering materials," *Sadhana: Academy Proceedings in Engineering Sciences*, Vol. 28, pp. 783–798.

26. Chatterjee, A. and Chakravorty, D. (1989), "Glass-metal nanocomposite synthesis by metal organic route," *Journal of Physics D: Applied Physics*, Vol. 22, pp. 1386–1392.

27. Harms, U., Shen, T.D., Schwarz, R.B. (2002), "Thermal conductivity of $Pd_{40}Ni_{40}\text{Å}_xCu_xP_{20}$ metallic glasses," *Materials Science and Technology*, Vol. 47, pp. 411–414.

28. Ashby, M.F. and Greer, A.L. (2006), "Metallic glasses as structural materials," *ScriptaMaterialia*, Vol. 54, pp. 321–326.

29. Albe, K., Ritter, Y., Şopu, D. (2013), "Enhancing the plasticity of metallic glasses: Shear band formation, nanocomposites and nanoglasses investigated by molecular dynamics simulations," *Mechanics of Materials*, Vol. 67, pp. 94–103.

30. Chatterjee, S., Saha, S.K., Chakravorty, D. (2016). "Glass–Based Nanocomposites" in *Glass Nanocomposites*, doi:10.1016/B978-0-323-39309-6.00002-X.

31. El-eskandarany, M.S. and Al-Hazza, A. (2017), "Fabrication of new metallic glassy $Ti_{40.6}Cu1_{5.4}Ni_{8.5}Al_{5.5}W_{30}$ alloy powders by mechanical alloying and subsequent SPS consolidation," *Advanced Powder Technology*, Vol. 28, pp. 814–819.

32. Pekarskaya, E., Löffler, J.F., Johnson, W.L. (2003), "Microstructural studies of crystallization of a Zr-based bulk metallic glass," *ActaMaterialia*, Vol. 51, pp. 4045–4057.

33. Loffler, J.F. (2003), "Bulk metallic glasses," *Intermetallics*, Vol. 11, pp. 529–540.

34. Bhattacharya, S. (2016), "Electrical transport properties of ion-conducting glass nanocomposites," *Glass Nanocomposites: Synthesis, Properties and Applications*, pp. 181–214.

35. Zhang, Z., Mao, S.X., Wang, J., Sheng, H., Zhong, L. (2014), "Formation of monatomic metallic glasses through ultrafast liquid quenching," *Nature*, Vol. 512, pp. 177–180.

36. Sharma, S., Chandra, R., Kumar, P., Kumar, N. (2015), "Thermo-mechanical characterization of multi-walled carbon nanotube reinforced polycarbonate composites: A molecular dynamics approach," *ComptesRendusMécanique*, Vol. 343, pp. 371–396.

37. Choi-Yim, H. and Johnson, W.L. (1997), "Bulk metallic glass matrix composites," *Applied Physics Letters*, Vol. 71, pp. 3808–3810.

38. Wang, Y.B., Xie, X.H., Li, H.F., Wang, X.L., Zhao, M.Z., Zhang, E.W., Qin, L. (2011), "Biodegradable CaMgZn bulk metallic glass for potential skeletal application," *ActaBiomaterialia*, Vol. 7, pp. 3196–3208.

39. Calin, M., Gebert, A., Ghinea, A.C., Gostin, P.F., Abdi, S., Mickel, C., Eckert, J. (2013), "Designing biocompatible Ti-based metallic glasses for implant applications," *Materials Science and Engineering C*, Vol. 33, pp. 875–883.

40. Murr, L.E., Inal, O.T., Wang, S.H. (1981), "Explosive shock deformation of metallic glasses," *Materials Science and Engineering*, Vol. 49, pp. 57–64.

41. Gloriant, T. (2003), "Microhardness and abrasive wear resistance of metallic glasses and nanostructured composite materials," *Journal of Non-Crystalline Solids*, Vol. 316, pp. 96–103.

42. Schroers, J., Kumar, G., Hodges, T.M., Chan, S., Kyriakides, T.R. (2009), "Metallic glass for biomedical applications," *The Journal of the Minerals, Metals & Materials Society*, Vol. 61, pp. 21–29.

43. Kobayashi, A., Yano, S., Kimura, H., Inoue, A. (2008), "Fe-based metallic glass coatings produced by smart plasma spraying process," *Materials Science and Engineering B: Solid-State Materials for Advanced Technology*, Vol. 148, pp. 110–113.

44. Siegrist, M.E. and Löffler, J.F. (2007), "Bulk metallic glass-graphite composites," *ScriptaMaterialia*, Vol. 56(12), pp. 1079–1082.

45. Li, J.C., Chen, X.W., Huang, F.L. (2018), "Ballistic performance of tungsten particle/metallic glass matrix composite long rod," *Defence Technology*, doi:10.1016/j.dt.2018.06.009.

46. Villapún, V.M., Zhang, H., Howden, C., Chow, L.C., Esat, F., Pérez, P., González, S. (2017), "Antimicrobial and wear performance of Cu-Zr-Al metallic glass composites," *Materials and Design*, Vol. 115, pp. 93–102.

47. Chang, C.M., Yang, C.J., Wang, K.K., Liu, J.K., Hsu, J.H., Huang, J.C. (2017), "On the reflectivity and antibacterial/antifungal responses of Al-Ni-Y optical thin film metallic glass composites," *Surface and Coatings Technology*, Vol. 327, 75–82.

48. Etiemble, A., Der Loughian, C., Apreutesei, M., Langlois, C., Cardinal, S., Pelletier, J.M., Steyer, P. (2017), "Innovative Zr-Cu-Ag thin film metallic glass deposed by magnetron PVD sputtering for antibacterial applications," *Journal of Alloys and Compounds*, Vol. 707, pp. 55–161.

49. Davidson, M., Roberts, S., Castro, G., Dillon, R.P., Kunz, A., Kozachkov, H., Hofmann, D.C. (2013), "Investigating amorphous metal composite architectures as spacecraft shielding," *Advanced Engineering Materials*, Vol. 15, pp. 27–33.

50. Singh, M. and Asthana, R. (2007), "Joining of zirconium diboride-based ultrahigh-temperature ceramic composites using metallic glass interlayers," *Materials Science and Engineering: A*, Vol. 460–461, pp. 153–162.

51. Pan, C.T., Chen, Y.C., Lin, P., Hsieh, C.C., Hsu, F.T., Lin, P., Huang, J.C. (2014), "Lens of controllable optical field with thin film metallic glasses for UV-LEDs," *Optics Express*, Vol. 22, pp. 1544–1553.

52. Pan, C.T., Chen, Y.C., Yang, T.L., Lin, P., Lin, P., Huang, J.C. (2015), "Study of reflection-typed LED surgical shadowless lamp with thin film Ag-based metallic glass," *Optik*, Vol. 127, pp. 2193–2196.

53. Chang, C.M., Wang, C.H., Hsu, J.H., Huang, J.C. (2014), "Al-Ni-Y-X (X = Cu, Ta, Zr) metallic glass composite thin films for broad-band uniform reflectivity," *Thin Solid Films*, Vol. 571, pp. 194–197.

54. Tian, R. and Qin, Z. (2014), "Bulk metallic glass $Zr_{55}Cu_{30}Al_{10}Ni_5$ bipolar plates for proton exchange membrane fuel cell," *Energy Conversion and Management*, Vol. 86, pp. 927–932.

55. Bian, Z., Wang, R.J., Wang, W.H., Zhang, T., Inoue, A. (2004), "Carbon-nanotube-reinforced Zr-based bulk metallic glass composites and their properties," *Advanced Functional Materials*, Vol. 14, pp. 55–63.

56. Tregilgas, J.H. (2004), "Amorphous titanium aluminide hinge," *Advanced Materials and Processes*, Vol. 162, pp. 40–41.

57. Liversidge, G.G, Bishop, J.F., Czekai, D.A. (1980), "Composite penetrator," *US Patent No.* US6010580A.

58. Gorsse, S., Chevalier, B., Orveillon, G. (2008), "Magnetocaloric effect and refrigeration capacity in $Gd_{60}Al_{10}Mn_{30}$ nanocomposite," *Applied Physics Letters*, Vol. 92, p. 122501.

59. Zhu, T.J., Yan, F., Zhao, X.B., Zhang, S.N., Chen, Y., Yang, S.H. (2007), "Preparation and thermoelectric properties of bulk in situ nanocomposites with amorphous/nano-crystalhybrid structure, *Journal of Physics D: Applied Physics*, Vol. 40, p. 6094.

60. Inoue, A. and Nishiyama, N. (2007), "New bulk metallic glasses for applications as magnetic-sensing, chemical, and structural materials," *MRS Bulletin*, Vol. 32, pp. 651–658.

5 Study of Damping Behavior of Metallic Glass Composites at Nanoscale Using Molecular Dynamics

Sumit Sharma, Prince Setia, Uday Krishna Ravella, and Gaurav Sharma

CONTENTS

5.1 INTRODUCTION

One of the most striking behaviors in the glass-forming materials is the observed change of many orders of magnitude of their dynamic properties (i.e., viscosity, relaxation time, etc.) without any apparent change of their structure [1]. The presence of an increased local order on approaching the glass transition has been suggested by simulations, but any direct experimental evidence of this ordering has been elusive to date [2]. It is maybe even more striking that the dynamic behavior is apparently identical in materials very different from a chemical point of view, such as metallic glasses and polymers, encompassing all different types of chemical interactions (ionic, covalent, hydrogen bonded, coulombic, etc.). However, it is not completely clear whether all observations found for more classical glass formers are valid for BMGs since the latter have been discovered more recently and are relatively less studied. One of the main methods of investigation of the dynamics is by relaxation measurements (i.e., dielectric relaxation and mechanical relaxation) [3–7]. Relaxation spectra of glass-forming materials are generally characterized by at least two types of relaxation: (i) the structural (also called α) relaxation, which is linked to the glass transition phenomenon and is observable near the glass transition, and (ii) the secondary relaxations (named β, γ, etc.), which are usually connected to the local atomic or molecular movements, and are more evident in the glassy state [8].

The mechanical relaxation characteristics of in-situ $Ti_{48}Zr_{20}V_{12}Cu_5Be_{15}$ metallic glass matrix composite with dendrites reinforced were investigated by Qiao et al. [9] using mechanical spectroscopy. An abnormal internal friction behavior below the glass transition temperature was detected. Considering the irreversible feature and based on the transmission electron microscopy analysis, the abnormal internal friction behavior was mainly induced by the precipitation of nanocrystals in the dendritic phase. In order to understand better the dynamic mechanical properties of the $Ti_{48}Zr_{20}V_{12}Cu_5Be_{15}$ metallic glass matrix composite, the kinetic characteristics of glass transition were analyzed in the framework of quasi-point defects theory. In addition, an isothermal annealing conducted at lower-than-glass-transition temperature induced a significant change on both modulus and loss factor; the kinetics of relaxation under annealing were well-described by the stretched exponential relaxation equation.

Wang et al. [10] reported the study of the interplay of α and slow β relaxations in strong and fragile bulk metallic glasses (BMGs) of $Pd_{40}Ni_{10}Cu_{30}P_{20}$, $La_{57.5}(Cu_{50}Ni_{50})_{25}Al_{17.5}$, $Pd_{40}Ni_{40}P_{20}$, and $(Cu_{50}Zr_{50})_{92}Al_8$ BMG-forming liquids using isothermal dynamic mechanical spectra. All dynamic mechanical measurements were performed on a TA DMA-2980 dynamic mechanical analyzer (DMA) by the single-cantilever bending method. A sinusoidal strain was applied during all dynamic mechanical measurements. The frequency dependence of Young modulus ($\pm 1\%$) was determined by the frequency sweep measurements. The temperature error was controlled within $\pm 0.1°C$ when isothermal measurements were performed. Before DMA measurements, all samples had been heated into their respective supercooled liquid regions and cooled at a constant rate (≤ 40 K/min) to room temperature in order to erase the interfering of the formation of the as-cast samples.

La-based BMGs have superior glass-forming ability, which can form a BMG over 30 mm in diameter by copper mold casting. To understand the structure and property relationship in such bulk samples is more important for industrial applications. Wang et al. [11] presented mechanical spectroscopy measurements on a $La_{62}Al_{14}(Cu_{5/6}Ag_{1/6})_{14}Ni_5Co_5$ BMG that had a critical size at least 20 mm in diameter and also under different annealing conditions to explore the relationship between the β relaxation and internal structure. In addition, the compressive properties of the as-cast and different pre-annealed samples in 2 mm diameter were examined. A correlation between β relaxation and mechanical properties of this BMG was demonstrated and discussed. It was found that the high temperature (below T_g) favors the β relaxation (excess wing) in the high frequency regime. Combined with DSC measurements, the β relaxation could be promoted by a short-time pre-annealing that only triggers atomic rearrangement in cages without new bond formation, corresponding to the disappearance of the exothermic event below T_g and no occurrence of enthalpy recovery above T_g. In contrast, no obvious correlation between the β relaxation and the compressive plasticity of this alloy at ambient temperature was observed.

Changes in intermittent shear avalanches during plastic deformation of a $Cu_{50}Zr_{45}Ti_5$ (atomic percent) alloy in response to variant structures, including fully glassy phase and/or nanocrystal/glass binary phase, were investigated by Wang et al. [12].

Second crystalline phases were introduced into the glassy-phase matrix of a $Cu_{50}Zr_{45}Ti_5$ metallic glass to interfere with the shear avalanche process, which could release the shear-strain concentration, and then tune the critically dynamic behavior of the shear avalanche. By combining microstructural observations of the nanocrystals with the statistical analysis of the corresponding deformation behavior, the authors determined the statistic distribution of shear-avalanche sizes during plastic deformation and established its dependence on the geometric distribution of nanocrystals. The scaling behavior of the distribution of shear-avalanche sizes followed a power-law relation accompanied by an exponentially decaying scaling function in the pure metallic glass and the metallic glass containing the small nanocrystals, which could be described by the mean field theory. The large shear-avalanche events were dominated by structural tuning parameters, i.e., the resistance of shear banding and the size and volume fraction of the second crystalline phase in metallic glasses.

Srivastava et al. [13] studied the relaxation dynamics of $Zr_{44}Cu_{40}Al_8Ag_8$ BMG using mechanical spectroscopy. This alloy manifested a clear signature of a slow β relaxation process below the glass transition temperature. The activation energy and the size of STZ was determined using mechanical spectroscopy. In addition, two glass transition processes competing in the supercooled liquid region were observed along with the β relaxation process. The possible mechanism for the relaxation processes was discussed. The main relaxation process obeys the VFT equation above the glass transition temperature. The correlation between the relaxation process below and above T_g was not well understood. This metallic glass composition may be a suitable candidate to give further insight into the mechanism of various relaxation processes.

It could also be noticed that lowering heating rate and frequencies in DMA gives opportunity to invoke a possible relaxation process which otherwise may not be apparent.

Controlled activation of flow units and in-situ characterization of mechanical properties in metallic glasses are facing challenges thus far. Li et al. [14] introduced vibrational loading through the nanoscale dynamic mechanical analysis technique to probe vibration-accelerated atomic level flow that played a crucial role in the mechanical behavior of metallic glasses. The intriguing finding was that high vibrational frequency induced deep indentation depth, prominent pop-in events on load-depth curves and low storage modulus, exhibiting a vibration-facilitated activation of flow units in $Pd_{40}Cu_{30}Ni_{10}P_{20}$ metallic glass. Theoretical analysis revealed that vibration-moderated activation time-scale accelerates the activation of flow units and is responsible for the above scenario.

The dynamic mechanical properties of the miscible $Cu_{50}Zr_{50}$ and immiscible $Cu_{50}Ag_{50}$ amorphous materials were investigated by Wang et al. [15] to explore the relationship between the deformation mechanism and the relaxation of glass through molecular dynamics simulation with the modified embedded atom method (MEAM). It was found that the mechanical hysteresis of $Cu_{50}Ag_{50}$ glass was more pronounced than that of $Cu_{50}Zr_{50}$ glass. The storage modulus decreased with increasing loading period or amplitude; while the loss modulus increased until the maximum, corresponding to the beginning of α relaxation. The β relaxation in both $Cu_{50}Zr_{50}$ and $Cu_{50}Ag_{50}$ glass showed excess tails in the loss modulus curves. However, the peak height on the left part in the curve of loss modulus as a function of temperature for the $Cu_{50}Ag_{50}$ glass was higher than that for the $Cu_{50}Zr_{50}$ system, which indicated that β relaxation in $Cu_{50}Ag_{50}$ glass was more likely to be activated than that in the $Cu_{50}Zr_{50}$ system due to a lower number of icosahedra-like clusters. The primary α relaxation always took place when the most probable atomic displacement reached a critical fraction (~23%) for $Cu_{50}Zr_{50}$ and (~21%) for $Cu_{50}Ag_{50}$ of the average inter-atomic distance, irrespective of whether the relaxation was induced by temperature (linear response) or by mechanic strain (non-linear). The fast atom was defined by the atom motion displacement to explore the dynamic heterogeneity of the glass. It was found that the internal fraction showed linearity with the number of fast atoms.

Many efforts have been devoted to clarify the origin of the mechanical relaxation of metallic glasses, but the nature of relaxation and the mechanical mechanism of metallic glasses are still elusive. These studies motivated the authors to investigate the role of nanofiller geometry in MG composites. While experimental research shows increments in damping properties of MG by adding CNTs, it is important to correlate this with the molecular dynamics (MD) study to explore the basic mechanisms of CNT-MG interaction during mechanical loading. It is also of particular importance to investigate how the single-layered graphene sheet (SLGS) influences the damping mechanisms of MGs at the atomic scale, which motivated this research by using MD simulations. This research will contribute to the design of MG-based nanocomposites with improved energy absorption capability.

To the best of the knowledge of the authors, this will be the first study to examine the energy absorption capability of CNTs and SLGs in MG matrix. This study utilized Biovia Materials Studio 7.0 to predict the damping properties of MG nanocomposites reinforced with unidirectional CNT and graphene. The remaining study is organized as follows. MD results are presented and discussed in Section 5.2. Finally, Section 5.3 concludes with a summary of the results obtained.

5.2 RESULTS AND DISCUSSION

Figure 5.1 shows the behavior of normalized storage modulus (E'/E_u) and loss modulus (E''/E_u) with temperature. The MD simulation was conducted at a constant driving frequency of 0.1 Hz and a heating rate of 5 K/min. The normalized parameter, E_u, is the unrelaxed modulus and approximately equals the value of the storage modulus at ambient temperature. The overall trend was similar to monolithic BMGs; there were differences in some regions resulting from the existence of embedded crystalline phase. The simulations were conducted with a constant heating rate of 5 K/min under different frequencies (i.e., 0.1, 0.3, and 3 Hz). Both the normalized storage modulus and loss modulus curves showed a similar trend to the polymers or monolithic BMGs [3]; the peak of E'/E_u at the onset of glass transition of $Cu_{64}Zr_{36}$ MG moved to higher temperature as the driving frequency was increased, where the amplitude of the movement depends on the activation energy of the corresponding moving units. As shown in Figure 5.1, an abnormal internal friction behavior occurred in the temperature range of 450–550 K, a significant "shoulder" was detected on the curve of loss modulus, appearing on the lower frequency curves (e.g., 0.1 or 0.3 Hz), which was not obvious with higher frequency. Intuitively, the abnormal "shoulder" was similar to the β relaxation

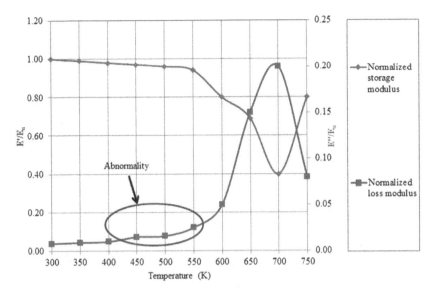

FIGURE 5.1   Normalized storage modulus E'/E_u and loss modulus E''/E_u of $Cu_{64}Zr_{36}$ MG as a function of temperature (T).

in some BMGs [16], which might be related to the intrinsic structural relaxation of amorphous materials. However, there was no such abnormal internal friction behavior found for the monolithic $Cu_{64}Zr_{36}$ BMGs [17], which means the abnormal internal friction behavior was most likely caused by the crystalline secondary phase.

Figure 5.2 shows the variation of normalized storage modulus (E'/E_u) and loss modulus (E''/E_u) of $Cu_{64}Zr_{36}$–CNT composite with temperature. It can be observed that, in comparison to the bulk MG, the normalized storage modulus for $Cu_{64}Zr_{36}$–CNT composite was higher (maximum by 12.5%). Also, the normalized damping loss modulus for $Cu_{64}Zr_{36}$–CNT composite was found to be lower (maximum difference of 6.67%) in comparison to bulk MG. This could be attributed to the higher damping capacity of the CNT reinforcement. CNTs are deformed together with the matrix if they are well-bonded. However, when the external load exceeds a critical value leading to de-bonding of CNTs from the matrix, CNTs will stop elongating together with the matrix and a further increase of load can only result in the deformation of the matrix. Thus, the matrix material starts to flow over the surface of the CNTs and deformation energy will be dissipated through the slippage between the CNTs and matrix. This phenomenon leads to the enhanced damping properties of nanocomposites compared to the bulk MG.

Figure 5.3 shows the variation of normalized storage modulus (E'/E_u) and loss modulus (E''/E_u) of $Cu_{64}Zr_{36}$–graphene composite with temperature. It can be observed that in comparison to the bulk MG, the normalized storage modulus for $Cu_{64}Zr_{36}$–graphene composite was higher (maximum by 25%). Also, the normalized damping loss modulus for $Cu_{64}Zr_{36}$–graphene composite was found to be lower (maximum by 30%) in comparison to bulk MG. Also, in comparison to the $Cu_{64}Zr_{36}$–CNT composite, the normalized storage modulus of $Cu_{64}Zr_{36}$–graphene composite

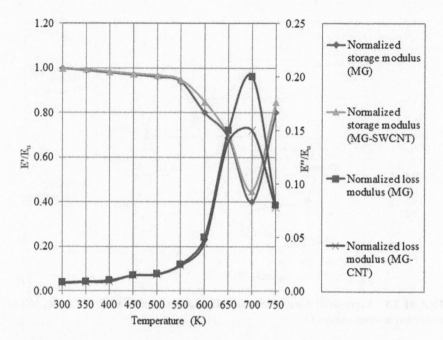

FIGURE 5.2   Normalized storage modulus E'/E_u and loss modulus E''/E_u of $Cu_{64}Zr_{36}$–CNT composite as a function of temperature (T).

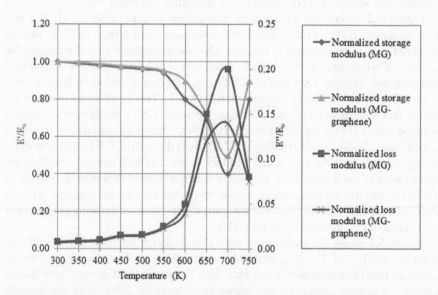

FIGURE 5.3   Normalized storage modulus E'/E_u and loss modulus E''/E_u of $Cu_{64}Zr_{36}$–graphene composite, as a function of temperature (T).

was found to be higher by approximately 11% at a temperature of 700 K. Similarly, the normalized damping loss modulus for $Cu_{64}Zr_{36}$–graphene composite was found to be 14% lower in comparison to the $Cu_{64}Zr_{36}$–CNT composite. This shows that graphene-reinforced MG composite has higher energy absorption capability in comparison to the CNT-reinforced MG composite.

5.3 CONCLUSIONS

MD simulations were performed to investigate the effect of CNT and graphene reinforcements on the damping properties of amorphous $Cu_{64}Zr_{36}$ MG. It was concluded that the CNTs can significantly enhance the storage modulus of amorphous MG, but graphene was found to be a better reinforcement. Some of the main conclusions that can be drawn based on the research conducted are highlighted below.

1. The peak of E'/E_u at the onset of glass transition of $Cu_{64}Zr_{36}$ MG moved to higher temperature as the driving frequency was increased.
2. An abnormal internal friction behavior occurred in the temperature range of 450–550 K; a significant "shoulder" was detected on the curve of loss modulus, appearing on the lower frequency curves (e.g., 0.1 or 0.3 Hz), which was not obvious with higher frequency.
3. In comparison to the bulk MG, the normalized storage modulus for $Cu_{64}Zr_{36}$–CNT composite was higher (maximum by 12.5%). Also, the normalized damping loss modulus for $Cu_{64}Zr_{36}$–CNT composite was found to be lower (maximum difference of 6.67%) in comparison to bulk MG.
4. In comparison to the bulk MG, the normalized storage modulus for $Cu_{64}Zr_{36}$–graphene composite was higher (maximum by 25%). Also, the normalized damping loss modulus for $Cu_{64}Zr_{36}$–graphene composite was found to be lower (maximum by 30%) in comparison to bulk MG.
5. In comparison to the $Cu_{64}Zr_{36}$–CNT composite, the normalized storage modulus of $Cu_{64}Zr_{36}$–graphene composite was found to be higher by approximately 11% at a temperature of 700 K. Similarly, the normalized damping loss modulus for $Cu_{64}Zr_{36}$–graphene composite was found to be 14% lower in comparison to the $Cu_{64}Zr_{36}$–CNT composite. This shows that graphene-reinforced MG composite has higher energy absorption capability in comparison to the CNT-reinforced MG composite.

The results of the study will help the researchers in the design and development of such MG-based nanocomposites that have high energy absorption capability.

ACKNOWLEDGEMENT

The authors state that this research has not been funded by any source and has received no funds from any institute or organization.

REFERENCES

1. Debenedetti, P.G. and Stillinger, F.H. (2001), "Supercooled liquids and the glass transition," *Nature*, Vol. 410, pp. 259–267.
2. Shintani, H. and Tanaka, H. (2006), "Frustration on the way to crystallization in glass," *Nature Physics*, Vol. 2, pp. 200–206.
3. Qiao, J.C. and Pelletier, J.M. (2014), "Dynamic mechanical relaxation in bulk metallic glasses: A review," *Journal of Materials Science and Technology*, Vol. 30, pp. 523–545.
4. Perez, J., Cavaillé, J.Y., Etienne, S., Fouquet, F., Guyot, F. (1983), "Internal friction in vitreous solids towards glass transition," *Annals of Physics*, Vol. 8, pp. 417–467.
5. Perez, J. (1988), "Defect diffusion model for volume and enthalpy recovery in amorphous polymers," *Polymer*, Vol. 29, pp. 483–489.
6. Pelletier, J.M., Louzguine-Luzgin, D.V., Li, S., Inoue, A. (2011), Elastic and viscoelastic properties of glassy, quasicrystalline and crystalline phases in $Zr_{65}Cu_5Ni_{10}Al_7.5Pd_{12.5}$ alloys," *Acta Materialia*, Vol. 59, pp. 2797–2806.
7. Perez, J. (2001), *Non-Crystalline Materials and the Science of Disorder*, Presses polytechniques et universitaires romandes, Lausanne, Switzerland.
8. Roland, C.M. (2011), *Viscoelastic Behaviour of Rubbery Materials*, New York, Oxford, 2011.
9. Lyu, G., Qiao, J., Gu, J., Min Song, M., Pelletier, J.M., Yao, Y. (2018), "Experimental analysis to the structural relaxation of $Ti_{48}Zr_{20}V_{12}Cu_5Be_{15}$ metallic glass matrix composite," *Journal of Alloys and Compounds*, Vol. 769, pp. 443–452.
10. Zhao, Z.F., Wen, P., Shek, C.H., Wang, W.H. (2010), "Relaxation behaviour on high frequency profile in strong/fragile metallic glass-forming systems," *Journal of Non-Crystalline Solids*, Vol. 356, pp. 1198–1200.
11. Wang, X.D., Liang, D.D., Ge, K., Cao, Q.P., Jiang, J.Z. (2014), "Annealing effect on beta-relaxation in a La-based bulk metallic glass," *Journal of Non-Crystalline Solids*, Vol. 383, pp. 97–101.
12. Tong, X., Wang, G., Yi, J., Ren, J.L., Pauly, S., Gao, Y.L., Zhai, Q.J., Mattern, N., Dahmen, K.A., Liaw, P.K., Eckert, J. (2016), "Shear avalanches in plastic deformation of a metallic glass composite," *International Journal of Plasticity*, Vol. 77, pp. 141–155.
13. Srivastava, A.P., Ştefanov, T., Srivastava, D., Browne, D.J. (2016), "Multiple relaxation processes in $Zr_{44}Cu_{40}Al_8Ag_8$ bulk metallic glass," *Materials Science & Engineering A*, Vol. 651, pp. 69–74.
14. Li, N., Liu, Z., Wang, X., Zhang, M. (2017) "Vibration-accelerated activation of flow units in a Pd-based bulk metallic glass," *Materials Science and Engineering A*, Vol. 692, pp. 62–66.
15. Wang, L., Wang, Y.Y., Peng, C.X., Li, X.L., Cheng, Y., Jia, L.J. (2018), "Dynamical mechanical analysis of metallic glass with and without miscibility gap," *Materials Science and Engineering: A*, Vol. 730, pp. 155–161.
16. Yu, H.B., Wang, W.H., Samwer, K. (2013), "The β relaxation in metallic glasses: An overview," *Materials Today*, Vol. 16, pp. 183–191.
17. Qiao, J.C., Pelletier, J.M., Kou, H.C., Zhou, X. (2012), "Modification of atomic mobility in a Ti-based bulk metallic glass by plastic deformation or thermal annealing," *Intermetallics*, Vol. 28, pp. 128–137.

6 MATLAB® Programming of Properties of Metallic Glasses and Their Nanocomposites

Raja Sekhar Dondapati

CONTENTS

6.1 INTRODUCTION

The discovery of amorphous structure in AuSi led to the discovery of metallic glass in 1960 [1]. Since then, widespread commercial and scientific interest has been gained by amorphous metals [2–4]. During the initial stage of its development, the primary objective of the scientific community was in the identification of under-cooled liquids, where nucleation crystals below glass transition temperature could be avoided [5,6]. Moreover, further research in the past two decades developed various alloy compositions, which allowed the prevention of reduction of critical cooling rates. Figure 6.1 shows the comparison between various materials based on their strengths. These alloys have been identified as bulk metallic glasses (BMGs), which can be fabricated at thicknesses greater than 1 mm and exhibit glass-forming ability.

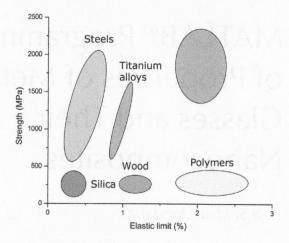

FIGURE 6.1 Comparison between various materials based on their strengths.

A wide range of applications for BMGs have been identified, starting in the mid-1990s; however, flaws in the material also emerged, such as low fracture toughness and lack of ductility [7,8]. By the end of the decade, BMGs received various world-wide research on BMGs identified vigorous funding. Impact on high precession and low thickness has been made by BMGs, while toughness, hardness, and high strength have been explored for various load applications, such as military vehicles, panels, and spacecraft shielding [9].

Metal alloys can be understood as amorphous phase, with access to infinite cooling rates because crystal growth and nucleation are dependent on time. A freestanding metal structure can be obtained by ribbon or splat quenching found to be achieving highest cooling rates are in the laboratory. However, for most metal alloys and metallic elements, crystal growth and nucleation intervene on a shorter time scale in the laboratory. Utilizing an induction coil and allowing the droplet to fall through a laser beam, which resulted in the splatting of droplets between two cooper paddles, allowed for achieving amorphous foils of 10–100 μm in thickness. Figure 6.2 shows the quenching of materials based on water.

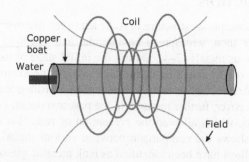

FIGURE 6.2 Quenching of materials based on water.

The fabrication of hundreds-of-meters-long strips can be made possible by the development of ribbon-quenching techniques. The ribbon-quenching technique consists of spraying molten alloy onto a spinning wheel, which allows the alloys to cool from one side by the wheel while the other side is free-cooled. Research in metallic glass for tuning toward the development of alloys, which could be formed in bulk dimension, took place by the 1980s, which was typically in the range of ~1 mm in diameter or thickness. The increase in critical cooling rates of the order of magnitude presented a major challenge in the formation of BMGs, e.g., 10^3 K/s in copper mold casting and 10^7 K/s in splat quenching. The best glass-forming system yet known is the alloy rods of up to 85 mm diameter, which exhibited the lowest known critical cooling rate of 0.067 K/s.

Dozens of alloy systems have been discovered in the past three decades, which can be formed into precious metals systems (Ag-, Au-, Pt-, and Pd-based), ferromagnetic alloy systems (Ni-, Fe-, and Co-based), and practical systems (Cu-, Ti-, and Zr-based). In the late 1990s, commercial applications for BMGs started in several Japanese ventures, and California demonstrated the promise of the new materials. Application ranges include springs, cell phone cases, golf clubs, sporting equipment, and biomedical implants, which were quickly developed by die-casting methods. The BMGs are ideal materials for applications typically held by polymers due to their unique chemistry. Figure 6.3 shows a schematic representation of the structure of quasicrystalline and crystalline structures. Regardless of the guarantee of supplanting plastics, similar to the early target, the global commercialization of BMGs did not occur due to a few unexpected drawbacks. It is currently realized that the mechanical properties of BMGs (especially fatigue limit and fracture toughness) are subjected to material quality, oxygen content, and the nearness of undesirable stages or incomplete crystallization. Besides, extensive scale die-casting, for the most part, includes melting crucibles and low-vacuum systems, which add contaminants and oxygen to BMGs. High-volume parts, produced using BMGs can be brittle if low-quality material is utilized. Regardless of these inadequacies, BMGs (of any quality or synthesis) are naturally fragile, which constrains any application where a section might be exposed to unconfined (tractable) loads. Attributable to an absence of microstructure to stop split arrangement, no solid metallic glass displays flexibility in strain, in spite of some composites that have substantial break durability. At present, there are currently correlations between thickness, crack durability, plastic zone

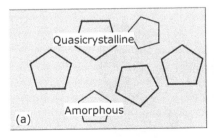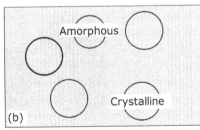

FIGURE 6.3 Schematic representation of the structure of (a) quasicrystalline and (b) crystalline structures.

size, and bending ductility, which exhibit that in plates or bars more than ~3 mm in thickness, BMGs fail in a brittle way [10–13]. In spite of the fact that these issues have been tended by making BMGs and planning applications for BMGs under 3 mm, the arrangement in the mid-2000s was to create BMG composites, where extra stages could be specifically added to BMGs to capture break spread.

By 2007, advancement was made in the rising parallel field of BMGMCs by building up a progression of rules for making toughened composites [14]. The composite framework needs an exceptionally processable solid metallic glass with crystallization. Zr-Be-based composites are ideal for BMGMC development due to the absence of stable mixes between the constituents and Be, which undergoes heterogeneous nucleation activated by the dendrites amid extinguishing. The composite framework needs a stable crystal that does not cause heterogeneous nucleation of the grid. These dendrites can be framed in harmonious two-stage frameworks, or by incompletely taking the shape of the BMG amid cooling. The shear modulus (G) of the dendrite should be lower than the grid. The microstructure ought to be coarsened to coordinate the incorporation size and dispersing with the plastic zone estimate (basic shear band length) of the BMG network. This standard was produced by observing a cooling rate reliance on the size of the dendrites in a BMGMC. At the point when cooled from the liquidus, time for nucleation and development is constrained, making a microstructure that fluctuates upon the nearby cooling rate. The real achievement was the understanding that toughness was identified with the span of these dendrites. This prompted the improvement of semisolid preparing, a transitional extinguishing step where the BMGMC is held simply over the solidus temperature in the two-stage locale and the dendrites are permitted to thermodynamically coarsen preceding the last extinguishing. Figure 6.4 shows the atomic structure of nanocrystalline material. This system demonstrated, to a great degree, making expansive coarsened microstructures, and malleability in BMGMCs was expanded to >10% while crack break toughness interface between the incorporation and the BMG network, with the end goal that split development does not continue along the limit of the considerations.

This rule was produced to clarify the absence of ductility in powder-reinforced ex-situ BMGMCs, even those where the size and creation of the dendrite coordinates in-situ composite. Oxide layers, or porosity, at the interface between the incorporations and the lattice, for the most part prompt splits proliferating along the limits.

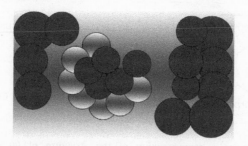

FIGURE 6.4   Atomic structure of nanocrystalline material.

TABLE 6.1

Composition of Glass Used for Electrodeposition of Silver (in mole%)

Glass no.	Li_2O	CaO	Al_2O_3	B_2O_3	SiO_2
1	20	13	3	4	60
2	27	13	3	4	53
3	30	12	3	...	55
4	35	10	3	...	53

Possessing short-range ordering, metallic glasses are amorphous in nature and have a nonperiodic arrangement of atoms in crystal structure [15]. Table 6.1 shows the composition of glass used for electrodeposition of silver (in mole%).

The distinction in the densities between the as-cast indistinct and completely solidified states is around 0.5%, which is a lot smaller than the recently revealed esteem scope of about 2% for the glasses with a lot higher R_c over 105 K/s. Such little contrasts in qualities show that mass indistinct amalgams have higher thickness arbitrarily stuffed nuclear arrangements. In spite of the fact that the distinction in densities between the indistinct and crystalline stages is little, the thickness increments in an efficient manner on toughening to basically loosened up glass and at last the precious stones. The strength of the supercooled fluid offers a bizarre chance to ponder the temperature reliance of consistency. Mass metallic glasses have extremely high return quality and high flexible farthest point. The Hookean flexible strain that a metallic glass can bolster in pressure or in twisting is practically twofold that of an industrially accessible crystalline material. Metallic glasses are chief spring material. They have high break quality combined with 2%–3% of versatile strain. Ordinary aluminium, titanium combinations, and steels can support 1%–2% of versatile strain. A vast area of high crack quality and versatile strain can be accomplished by nanoquasicystallization and by strengthening clay particles into the formless framework. Metallic glasses are moderately feeble in shear, so they have a similarly high Poisson's proportion, which shifts between 0.35 and 0.37. After yielding, a metallic glass will in general create a confined shear band that navigates the sample [16].

Bulk metallic glasses (BMG) show especially appealing mechanical properties and huge flexible strains (up to 2%). Among nebulous amalgams, Mg-based BMG are of uncommon combinations for structural applications. Fracture stress is fundamentally upgraded by the Fe expansion, which might be most likely identified with the cooperation between the dendrites and proliferating shear groups through the glass. Nanocrystallization in the contemplated Mg–Cu–Gd indistinct combination results in a lessening of the crack pressure, which contrasts from results detailed in different BMGs. The evident particularity of the Mg-based BMGs might be identified with the qualities of the crystal populace or their level of unwinding of the glass before deformation [17].

At the point when a ferromagnetic Fe–(Al, Ga) metalloid mass smooth combination was incorporated for the first time, Fe-based metallic glasses have pulled in colossal enthusiasm for the logical network, for the most part because of their great delicate attractive properties. Contrasted with other mass metallic glasses (BMGs), (Fe–Co)-based formless combinations are likewise especially alluring for building applications because of their blend of ultrahigh quality strength. Besides, nanocomposites derived from halfway devitrification of polished amalgams ordinarily show better quality and pliancy than the flawless solid BMGs. From a tribological perspective, it has been demonstrated that the wear resistance is likewise improved by nanometer-scale scattering of crystallites or semicrystallites in the undefined grid. It is vital that in spite of the fact that the attractive properties of smooth and solidified Fe–Co–Si–B–Nb amalgams have been as of late explored, few investigations have provided details regarding the mechanical properties (hardness, Young's modulus, or wear obstruction) of this framework. The impact of thermally incited devitrification on the flexible and mechanical properties of $Fe_{36}Co_{36}B_{19.2}Si_{4.8}Nb_4$ mass metallic glass has been explored. Our outcomes uncover that $Fe_{36}Co_{36}B_{19.2}Si_{4.8}Nb_4$ mass shiny combination shows ultrahigh hardness, high Young's modulus, and great wear obstruction. The high quality of the as-cast example is fundamentally related to the solid holding between the constituent components, especially in the nuclear sets containing Nb and Si. The hardness and versatile properties are upgraded by strengthening medicines while the Poisson's proportion is found to logically diminish. In the example toughened underneath Tg, mechanical solidifying is ascribed to the precipitation of semicrystalline particles and the development of the metastable complex-cubic stage $Fe_{23}B_6$, which is known to have vast hardness. For the partly and completely crystalline examples, the upgrade on hardness is credited to other extra components, for example, strong arrangement, solidifying, or the cooperation between separations or, furthermore, twin limits and different deformities. Concerning wear properties, the best wear opposition is the example demonstrating a structure comprising nanocrystals implanted in a formless grid. The different microstructural mechanisms are responsible for these annealing-induced changes in mechanical and elastic properties [18].

The magnetocaloric effect (MCE) is defined as comprises in the temperature change of attractive materials under the utilization of an outside magnetic field. This MCE is the key impact whereupon is based the very encouraging and eco-accommodating attractive refrigeration innovation. $Gd_{60}Mn_{30}In_{10}$ and $Gd_{60}Mn_{30}Ga_{10}$ nanocomposites have been arranged by single-roller melt turning. The magnetocaloric impact and refrigeration limit of $Gd_{60}Mn_{30}Ga_{10}$ and $Gd_{60}Mn_{30}In_{10}$ melt spun nanocomposites were examined. It was discovered that the nanocystallites shaped in the formless grid are diverse between melt spun $Gd_{60}Mn_{30}Ga_{10}$ and $Gd_{60}Mn_{30}In_{10}$ tests [19].

The advancement of mass metallic glasses and composites for improving the mechanical properties has happened with the revelation of numerous pliable metallic glasses and glass lattice composites with second stage scatterings with various length scales. The significant favorable position of metallic glasses is their high versatile strain ($\sim$2%), which is a lot higher than that of normal crystalline metallic composites (<1%). Also, because of the absence of microstructural highlights, for example, grains, grain limits, and disengagements, the corrosion resistance and mechanical quality of mass metallic glasses are high. The mechanical properties of solid BMGs are an interesting mix of high quality, from 1 GPa in the instance of Mg-based BMGs up to 3–4 GPa on account of Fe- or Co-based BMGs, and low Young's modulus (80–90 GPa), together with a versatile strain of around 2%. Be that as it may, the naturally visible flexibility of metallic glasses is somewhat low. The elements overseeing the strength and ductility and the toughening systems in BMG composite microstructures incorporate the properties of the fortifying stages, for example, their flexible properties; their yield quality and flexibility for upgrading the load exchange from the lattice to the fortifying phase(s); and the properties of the interface between the second stage and the shiny framework as far as holding quality and wettability, together with the volume part, size, and morphology of the second stages. The wide scope of unrivaled mechanical properties of bulk metallic glasses and their composites renders an extraordinary open door for growing new propelled materials, which can be utilized for an assortment of design applications. The exceptional structure–property relationship of metallic glasses is of logical as well as mechanical intrigue. Countless grid composites with quasi-crystalline, crystalline, and noncrystalline second stage scatterings on the nanometer-micrometer length scale have been delivered either by cementing, mechanical alloying and union, or an auxiliary treatment (e.g., strengthening or serious plastic distortion) of undefined forerunners. Tailorable mechanical properties (quality, hardness, flexibility, durability) can be accomplished from either in-situ or ex-situ BMG composites by receiving legitimate handling procedures and tuned-in pieces to change the size and the volume division of the second stage scatterings [20].

Recently, there has been significant logical and mechanical enthusiasm for an assortment of metal grid composites as an approach to improve mechanical properties contrasted with unreinforced alloys. Those materials are made by strengthening amalgams with long or short strands, hairs, or particles. Ceaselessly strengthened composites are having extreme quality with high hardness and are anisotropic in nature. Discontinuously strengthened metal network composites have been shown to offer basically isotropic properties with generous enhancements in quality and firmness in respect to those accessible with unreinforced materials [21].

Another class of ductile β stage fortified mass metallic glass composites were made by an effectively attainable in-situ preparing strategy from a homogeneous $Zr_{56.2}Ti_{13.8}Nb_{5.0}Cu_{6.9}Ni_{5.6}Be_{12.5}$ liquid. The microstructure of the subsequent two-stage material comprises a dendritic Zr– Ti rich β stage with a body focused cubic structure, which is installed in a completely indistinct lattice. The two stages are in compound balance and, furthermore, show molecularly sharp, cozy, and evidently solid interfaces. The framework stage has fundamentally the same organization as Vitreloy 1 and displays in solid structure mechanical properties commonplace for a bulk metallic glass. The fortifying β stage appears in solid structure as a delicate and pliable mechanical conduct with articulated work solidifying and a serrated stream conduct. The two stages consolidated in one material yield a composite with fantastic mechanical properties. Contrasted with Vitreloy, the best mass metallic glass known so far in the Zr-based framework, sway strength improved by a factor of 2.5 and normal malleable strains by a factor of 2.7. The yield point and work solidifying conduct of the composites can be customized by the handling conditions.

An extreme rigidity up to 1500 MPa was attained. The improved ductility and toughening of the composites can be clarified as far as the distortion component, which is overwhelmed by the β stage dendrites [22].

Homogeneous bulk metallic glasses (BMGs) demonstrate a progression of promising properties, e.g., high mechanical strength, broadened elastic and high hardness. As opposed to this, the plastic deformability of solid BMGs is somewhat poor as a rule, which limits essentially specialized applications. It has been demonstrated that the presentation of nanoparticles into BMGs can improve their ductility impressively. Molecular dynamics (MD) contemplates that relationships exist between the elastic moduli of a BMG and its plastic conduct. It is additionally realized that the powerful flexible properties change basically upon fractional crystallization of BMGs. In general, the forecast of properties and property blends of composite materials has been an essential undertaking. In an ongoing survey, the current hypothetical models and computational techniques to ponder compelling properties of complex composite materials have been gathered. While computational techniques turn out to be increasingly more vital for structure and enhancement of multiphase materials with appropriately reproducible microstructure, hypothetical models are successful for the estimation of patterns for a few classes of composite materials [23].

Since the revelation of Pd- and Zr-based bulk metallic glassy combinations (BMGs), there has been developing enthusiasm for hunting down compositions having high glass-forming ability. The Fe-based combinations are specifically noteworthy in view of their potential attractive applications. A most extreme breadth of 6 mm for formless bar tests was accounted for the alloy $Fe_{61}Co_7Zr_{10}Mo_5W_2B_{15}$. However, paramagnetic properties at room temperature, coming about because of the aggregate centralization of metalloids and nonmagnetic recalcitrant metals, especially the last components, limit its applications. Much better attractive properties have been accounted for Fe–Ga–Cr–Mo–P–B–C composites; however, the basic measurement for formless pole tests is restricted to 1 mm [9].

$[(Fe_{0.5}Co_{0.5})_{0.75}Si_{0.05}B_{0.25}]_{96}Nb_4$ glassy samples were produced using the injection copper mold casting technique. The samples show soft magnetic properties and a good thermal stability. The amorphous samples may have a kind of pre-defined SRO, clusters or nuclei, ready to crystallize, as indicated by the Avrami exponent $n{\sim}1.43$ and the very small incubation time. The energy barrier for crystallization is high, but once the right temperature is attained the crystallization proceeds immediately. The crystallization is diffusion controlled and needs high temperatures to start. Furthermore, transmission electron microscopy studies are required in order to rule out the exact structure of the sample. The magnetic properties follow the structure. The pre-existing nuclei do not influence the magnetic properties of as-cast samples. Such kinds of glassy samples may be good candidates for nanocrystallization studies upon small additions of elements which have positive heat of mixing with the main ingredients Fe and Co [10].

Metallic materials, which are generally utilized in auxiliary designing, have pulled in much consideration, especially for their mechanical properties. Research has unavoidably focused on accomplishing high strength and great plasticity. Decreasing the grain estimate has turned out to be an effective method for improving the quality of traditional crystalline materials, coming about in ultrafine-grained and nanostructured materials. Decreasing the grain size to the least physical breaking point in a confused nuclear game plan results in the arrangement of basic metallic glasses (MGs). This achievement in the field of material science invigorated far-reaching interest and another exploration network. From that point, materials quality was significantly upgraded consistently. Be that as it may, upgrading strength regularly prompts the lessening in plasticity. Along these lines, improving plasticity of MGs has, as of late, turned into another hot point. Accomplishing the ideal microstructure by in-situ planning is feasible. For instance, Co-based bulk metallic glasses (BMGs), having extraordinary quality as well as high innate fragility, can show a level of versatility by in-situ shaping dendrite stages. Superplastic metallic glass was accomplished at room temperature by directing the microstructure to give diverse hardness. Moreover, plasticity has been improved, to some extent, by adjusting the microstructure during casting. Surface mechanical attrition treatment (SMAT) is a newly developed technique. It has proven to be an effective method to improve the surface so as to significantly enhance the global mechanical properties of crystalline materials by means of repeated multidirectional impacts at high energy and high strain rate on the sample surface, resulting in severe plastic deformation. This is a flexible process to achieve specific structure and property requirements without changing the chemical composition. An epic auxiliary gradient metallic glass composite was framed by exposing a completely smooth combination to surface mechanical wearing down treatment. The basic slope comprised submicron-scale crystallites on the best surface, trailed by nanoscale crystallites and, furthermore, a completely shapeless glass lattice inside. Plasticity was extraordinarily improved by up to about 400%, contrasting that of as-cast metallic glass. Sub-framing crystallites are in charge of the improvement in plasticity [11].

The impact of installed nanocrystals on the mechanical properties of BMG composites was researched by compressive testing. At the point when nanocrystal scatterings were created, zirconium-copper-based BMGs that at first demonstrated no plastic disfigurement before break showed flexible conduct in pressure with about 10% deformation. Bulk metallic glasses (BMGs) have intriguing mechanical properties, for example, high strength up to 5 GPa, high elastic strain of about 2%, and numerous others extra alluring properties. Nonetheless, BMGs show the versatile locale bomb disastrously on one predominant shear band and show minimal plainly visible plasticity in a clearly fragile way. Plastic deformation in MGs at room temperature starts with initiation of shear change zones (STZs) in which groups of particles modify under a connected pressure. As the pressure is expanded, more STZs end up dynamic and associate with structure shear groups in which the plastic strain is limited. However, there have recently been several reports of significant macroscopic plastic deformation in CuZr-based metallic glasses containing nanocrystals and a Pt-based glass. The binary CuZr glass was repeatedly found to undergo plastic deformation of more than 50% in uniaxial compression [12].

Glass-metal nanocomposites including silver have particles inside traded lithia silicate glasses by an electro-deposition procedure. The silver molecule distances across run from 4.0 to 12.0 nm, contingent upon the salt particle focus in the antecedent glass. A large portion of the examples display metallic conduct. Be that as it may, the viable Debye temperature describing the resistivity variety is found to diminish radically as the molecule estimate is diminished from 6.0 to 4.0 nm. This emerges because of a bigger portion of atoms living at the surface of the particles. Nanoparticles of silver, copper, and iron, separately, with breadths going from 6.6 to 11.6 nm have additionally been developed inside a glass-clay by an ion exchange and reduction procedure. The electrical resistivity shows a temperature subordinate initiation vitality. The information can't, nonetheless, be fitted to either a $T^{-1/2}$ or $T^{-1/4}$ law. The activation energy in the temperature range 200–300 K is constrained by an electron burrowing system between the metal grains. In the lower temperature range, a quantum estimate impact seems, by all accounts, to be usable, descending to a low activation energy of about a couple meV [13].

Glass-metal nanocomposites, including copper and nickel, individually, have been combined in mass structure by hot squeezing sol-gel inferred silica-metal nanoparticles composite powders. The molecule distances across range from 9 to 17.5 nm. The examples display the trademark conduct of the metallic species in their nanocrystalline shapes. It is conceivable to ponder the electrical conduction conduct of a metal in the nanocrystalline state on the grounds that the permeation arrangement was accomplished in the framework. Be that as it may, this nanocomposite structure was restricted to a couple of micron thickness on the outside of the glass-fired concerned. Such nanocomposite structures with metallic permeation have additionally been created with a couple of micron thickness on an appropriate glass substrate utilizing a sol-gel strategy treatment. The sol-gel strategy was prior misused to get ready clay metal composites, diphasic aerogels [14].

6.2 METALLIC GLASS

Over the past few decades, numerous investigations have been carried out with metallic glasses owing to their properties and technological potential [24–27]. During the initial period, a high cooling rate of the order ~105 K/s persisted for the formation of glassy phase. However, very slow cooling rates are implemented for the synthesizing of bulk metallic glass (BMG). Fundamental interest in the properties of the disordered state of material have driven an immense research activity. Metallic glasses shows high elastic limit and yield strength compared to crystalline titanium and steel alloys. Coupled with 2%–3% of elastic strain, BMGs have high fracture strength. Moreover, upon nanocrystallization into amorphous matrix, a higher degree of elastic strain and fracture strength can be achieved.

Metallic glasses exhibit some attractive properties, which can be further improved using heat treatment processes. Furthermore, by the addition of ceramic material as a secondary phase, structural properties can be enhanced, such as ductility. Such composite material finds application in aircrafts, automobiles, and medical implants. Also, the machinability of these materials is high due to the large supercooled liquid regions. However, localized plastic deformation at room temperature occurs due to the formation of shear bands. Hence, the atomic structure and mechanical behavior of BMGs and their composites are not clearly understood.

6.2.1 NANOCOMPOSITES

Metallic glasses and quasicrystals forming alloys have favorable properties for many potential applications. However, bulk quasicrystals are mostly brittle, which can be removed by controlled crystallization of glass nanocrystal composites, such as $Zr_{69.5}Al_{7.5}Cu_{12}Ni_{11}$. The controlling of the formation of the nanostructure is of extreme importance since it determines the material properties. It has been found that Zr- and Ti-based nanocomposites have significantly high hardness (610 and 755 VHN, respectively). Misra et al. [28] studied the deformation in the nanostructure of bulk glass composites. A decrease in bulk modulus, Poisson's ratio, and specific volume were observed.

Metallic materials are generally crystalline in nature which possess translational and orientation periodic repetitiveness of their constitutive atoms in three-dimensional fashion, which are symmetric in nature. The concept changed completely when metallic glasses (MGs) and quasicrystals (QCs) were explored. Metallic glasses are amorphous and exhibit short-range ordering. Metallic glasses also possess nanoquasicrystalline phase when annealed, which exhibit significant potential to be used in various potential applications. At the early stage of discovering metallic glasses, there was a requirement of cooling the material at a very fast rate (typically 10^5 to 10^6 K/s) in order to form glassy, phase of the metal. But at a later stage, it was observed that blending several metals, which mix with each other freely but cannot easily crystallize together as they have different atomic sizes and structures, requires less cooling rate in order to form glassy phase, which is named bulk metallic glass. It has exhibited excellent properties (very high hardness and other mechanical properties like metals and forming abilities like polymers). But in spite of their strength, BMGs are not yet tough enough for load bearing applications, i.e., when the stress becomes very high, they can fracture without warning. This is due to very little global room temperature plasticity, and they deform by highly localized shear banding, thus showing brittleness in their stress-strain curve and leading to low toughness. This is known as quasibrittle deformation behavior. This can be improved by nanocrystallization of the material.

Eckert et al. [29] experimentally investigated the mechanical properties of $Zr_{55}Al_{10}Cu_{30}Ni_5$ BMG alloy with up to 17.5% of W nanoparticles dispersed homogeneously. The peaks in the XRD patterns are for the dispersed phase of the nanoparticles, and it is increasing with the volume fraction of the dispersed phase. W particles are homogeneously dispersed in the metallic glass matrix with the size distribution range of the nanoparticles of 35 ± 8 nm. The results of isothermal annealing show that the incubation behavior of crystallization of the BMG composite is not affected by the presence of the nanocrystals, but there is an acceleration in transformation into the crystalline state. The Vickers hardness of the material increased with the addition of the W nanoparticles as compared to the homogeneous metallic glass. This could be due to the overall contribution of both the phases for the hardness of the composite material. The increase in room temperature viscosity was also found. All the results indicate an increase in tensile/compression stress for the composite sample.

In another work [30], they experimentally investigated the mechanical properties of three samples of bulk metallic glass. They prepared three compositions of master alloys of BMG, e.g., $Cu_{47}Ti_{34}Zr_{11}Ni_8$ (Sample A), $Cu_{47}Ti_{33}Zr_{11}Ni_8Fe_1$

(Sample B), and $Cu_{47}Ti_{33}Zr_{11}Ni_8Si_1$ (Sample C), prepared by arc melting in a Ti-gettered argon atmosphere in a copper mold casting. XRD patterns show that no crystalline peaks were found, which indicates fully amorphous phase of all three alloys. It can be observed that, for all three alloys, compression stress exceeds 2000 MPa and Young's moduli exists between 100 and 109 GPa. Only Si-contained Sample C shows strain hardening and a plastic strain of about 4.4% and a high elastic strain limit of 2.2%, which indicates good room temperature ductility. Bright field TEM micrographs and electron diffraction patterns show that the Si-contained alloy (Sample C) is not fully amorphous, but contains 8–15 nm large FCC nanocrystals homogeneously.

This is the main reason for the improvement of the global plasticity as it forms multiple shear bands, leading to the improvement of the resistance against catastrophic crack propagation. This type of in-situ processing of BMG nanocomposites can be used as excellent structural material for its high strength and ductility.

Branagan et al. [31] investigated the superplasticity phenomenon in an iron-based BMG nanocomposite. The alloy considered in this work has an atomic stoichiometry of $(Fe_{0.8}Cr_{0.2})_{79}B_{17}W_2C_2$. A novel method was used to develop nanocomposite microstructures in bulk glassy iron-based materials by utilizing the self-assembling phenomenon in solid state transformations. A very dense and temperature-stabiliized microstructure was achieved, which is a challenging factor. It was also found that refinement occurs by utilizing this approach for the size and distribution of the secondary phase over the primary phase. This stabilized microstructured material exhibited tensile elongation of 230% and an ultimate tensile strength of 1800 MPa at 750°C. The strain rate sensitivity factor was found to be 0.51, which is very good evidence of the superplasticity phenomenon.

Amir Hossein Taghvaei et al. [32] investigated microstructure and corrosion resistance behavior of Ni-based nanocomposite coating reinforced with BMG particles. Metallic-glass particles used in this work were $Ni_{60}Cr_{10}Ta_{10}P_{16}B_4$ for reinforcement. The coating was applied by deposition through a DC electrodeposition in a typical Watt's bath. The average thickness of the coating was 60 μm with uniform distribution of the metallic glass particles over a steel substrate with less porosity and cracks at the interface. It was found that the maximum volume concentration of the MG particles was 6.4% for the particle concentration in the bath at 10 g/L. In this concentration maximum, microhardness was observed, which was found to be 137 HV, which is higher than the same of pure Ni coating. Corrosion improvement was found to be 45 times better than the pure Ni coating at 10 g/L particle concentration in the bath. The coating is uniformly distributed throughout the steel substrate, and there is no porosity and crack identified at the interface, and homogeneous distribution of the glassy particles with the average range of 4–8 μm can be found. At lower concentrations and at higher concentrations, heterogeneity can be found and agglomeration occurs at very high concentration. The corrosion current density initially decreases significantly from pure Ni coating to particle concentration of 10 g/L of the nanocomposite coating and again it increases up to the concentration of the 40 g/L. Thus it was found from their work that the optimum concentration of the MG particles in the Ni-based matrix is 10 g/L with excellent corrosion resistant behavior and enhanced microhardness.

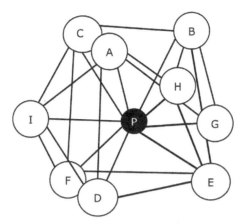

FIGURE 6.5 Schematic illustration of a trigonal prism capped with three half-octahedra consisting of Pd, Ni, and P atoms and a tetragonal dodecahedron consisting of Pd, Cu, and P atoms.

Branagan et al. [33] investigated the wear resistance characteristics of $Fe_{63}Cr_8Mo_2B_{17}C_5Si_1Al_4$ nanocomposite steel coating at a low cooling rate (10^4 K/s) for the formation of the amorphous metallic glass. For the coating, two methods were utilized (i.e., plasma spraying and high velocity oxy fuel spraying technique). It was found from their work that higher hardness could be found compared to conventional steels for both fully amorphous (10.2–10.7 GPa) and devitrified nanocomposite coatings (11.4–12.8 GPa). Excellent adhesion strength of the coating was found over a wide variety of metallic substrates and very good abrasion resistance quality was also found. Figure 6.5 shows a schematic illustration of a trigonal prism capped with three half-octahedra consisting of Pd, Ni, and P atoms and a tetragonal dodecahedron consisting of Pd, Cu, and P atoms. Figure 6.6 shows the process involved in the growth and formation of amorphous materials.

Gorsse et al. [34] studied the refrigeration capacity and magnetocaloric effect of a BMG nanocomposite. They studied the magnetic properties and magnetocaloric effect of Gd-based glass with a nominal composition $Gd_{60}Al_{10}Mn_{30}$ containing nanocrystalline of Gd. The nanocrystalline materials were of the diameter of less than 10 nm and the XRD pattern confirms the mixture of the crystalline material dispersed

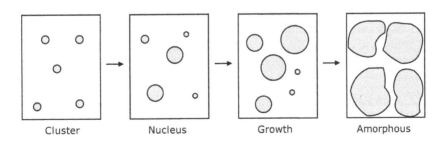

FIGURE 6.6 Process involved in the growth and formation of amorphous materials.

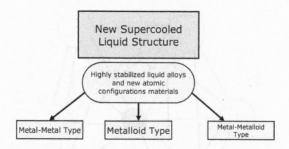

FIGURE 6.7   Division of supercooled liquid structure.

on amorphous material. Above 300 K temperature, the behavior of the nanocomposite can be represented by Curie-Weiss law with an effective magnetic moment of 5.8 μ_B and paramagnetic Curie temperature of 238 K. It also shows high refrigeration capacity and soft magnetic properties with wide temperature operability. Hence, it shows a very good magnetic refrigerant character at 150 K temperature, which is very challenging. Figure 6.7 shows the division of supercooled liquid structure.

Zhu et al. [35] investigated the thermoelectric properties of $Ge_{20}Te_{80}$ bulk in-situ nanocomposite of amorphous/nanocrystalline hybrid structure. The material was then heat-sealed by quartz tubes at 10^{-3}Pa. These sealed tubes were heated at 1173 K and then quenched in liquid nitrogen and thus amorphous bulk metallic structure was formed. The formation of nanocrystals was done in-situ by annealing between the glass transition temperature and crystallization temperature of the amorphous material. A fully amorphous structure can be found in the case of liquid nitrogen–quenched material as for the rapid cooling, but in the case of water-quenched material is fully crystalline in structure as for enough time to be crystallized for lower cooling rate. After annealing at 443 K for 2 hours, nanocomposite structure can be observed. The same can be observed in the case of HRTEM micrographs. The highest electrical conductivity can be found for crystallized material and of the nanocomposite material annealed at 443 K, and for both the cases it is almost constant regardless of temperature change (slightly decreasing trend can be found in the case of crystalline material). But in the case of amorphous material materials annealed at lower temperature, the electrical conductivity increases with the increase in temperature. The Seebeck coefficient varies with the temperature for different materials, and the highest can be found in the case of amorphous material; there is a decreasing trend with the increase of annealing temperature, and the lowest is for the crystalline structure. The highest power factor can be found for crystalline material, and material annealed at 443 K for 2 hours is next to that. So, it can be observed that this type of bulk metallic glass nanocomposite can be used as an excellent thermoelectric material.

6.2.2   Mechanical Properties

The difference in density between the fully crystallized state and the cast amorphous state is ~0.5%, which is much higher R_c above 10^5 K/s and reportedly much smaller than the previous range value of ~2% for the glasses. Hence, such small difference in density indicates the presence of randomly packed atomic configuration in bulk amorphous alloys. However, during the annealing process, the density increases in a systematic fashion, resulting in a stress-free structure, even though a small density difference exists between crystalline and amorphous phase. Moreover, an opportunity to analyze the temperature-dependent viscosity is offered by the stability of the supercooled liquid. Johnson et al. [36] conducted a study on bulk metallic glass (BMG) showing that viscosity fits very well with the Vogel-Fulcher equation over fifteen orders of magnitude, given by

$$\eta = \eta_0 \exp\left(\frac{DT_0}{T - T_0}\right) \tag{6.1}$$

where, T_0 represents the Vogel-Fulcher temperature and D represents the fragility index. Angell [37] distinguished between the strong and fragile glasses. Based upon such characterization, it was found that most metallic glasses are fragile and some organic polymers and silica silicates are strong glass. However, the value of D ranges from 20 to 25 for strong BMGs. Above Tg, viscosity of many orders of magnitude is found in a strong glass. Under a viscosity high in magnitude, retardation of nucleation tends to occur. However, strong glasses are resistant to crystallization. However, based on energy levels, these glasses are closer to crystals. A strong mega basin is exhibited by theses crystals for glass transition. Bending or tension can be supported by metallic glass under the Hookean elastic strain, which is double compared to the commercial crystalline structures.

The recent interest is nanocrystallization induced by deformation in amorphous alloys. In such material, deformation is highly localized in nature and strain hardening is not shown by them. Figure 6.8 shows the shear action of materials under external loading conditions. Adiabatic heating, strain softening, and formation of shear bands is observed during localized deformation. During nanoindentation in bulk metallic glass, nanocrystallization takes place in experimentation [38]. When the stress is highest, bands discontinuous in nature are formed; nanoparticles of Zr_2Ni are observed at the sides and tip of indentation. Such crystallization is due to the enhancement of mobility of atoms under high-stress conditions. However, enhancement of properties also takes place upon partial crystallization [5].

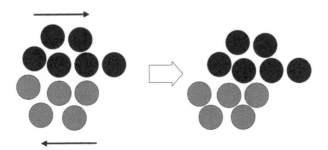

FIGURE 6.8 Shear action of materials under external loading conditions.

The microstructure controlling involves parameters, such as volume fraction, precipitated phase, and interface shape. Moreover, reinforcements hinder the propagation of shear bands, improving the mechanical property. Attractive properties are demonstrated by BMGs, which are useful in diverse industries. Moreover, BMGs have excellent processability and machinability. Viscous flow formation takes place below crystallization temperature, which makes them a suitable candidate for shape-forming processes.

6.2.3 CHEMICAL PROPERTIES

High order of corrosion resistance is offered by metallic glass. The chemical homogeneity and the absence of defects allow for the high corrosion resistance. Moreover, an important role is played by the composition of glasses. As revealed by the studies on melt-spun ribbons by Raja et al. [39], Mo, Cr, and P can cause enhancement of the corrosion resistance. Since BMGs have possible applications as structural materials, corrosion resistance is a more critical phenomenon. The influence of chemical composition conducted by Inoue (2000), with Ta– and Nb–amorphous alloys demonstrate high resistance to corrosion in NaCl and HCl solutions. Figure 6.9 shows the comparison between the existence of fuel particles in (a) agglomerated form and (b) dispersed form.

6.2.4 MAGNETIC PROPERTIES

Soft ferromagnetic materials have tunable nanocomposite microstructure, which have potential applications in the development of magnetic devices and sensors [40]. One such material is the cobalt-based metallic glass (CoMG), which possesses exceptional ultralow coercivity (~1 A/m), soft ferromagnetic properties, low saturation magnetostriction ($\lambda_s = 10^{-7}$ ppm), and low remnant magnetization (M_r) [41]. Moreover, they exhibit high Curie temperature [42] and excellent mechanical properties [43]. With a homogenous isotropic structure along with a lack of grain boundaries in metallic glasses, low hysteresis losses and high saturation magnetization make

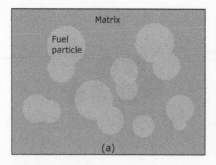 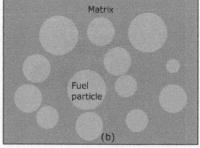

FIGURE 6.9 Comparison between the existence of fuel particles in (a) agglomerated form and (b) dispersed form.

them reliable for soft ferromagnetic applications, such as fluxgate sensors and electromagnetic shielding [44,45]. When $H_c > 10$ A/m, materials are called hard ferromagnetic materials whereas $H_c < 10$ A/m corresponds to the classification called soft ferromagnetic [46]. To date, numerous nanocrystalline alloys have been developed, such as NANOPERM, FINEMET, and HITPERM, for addressing the challenges in soft ferromagnetic materials [47,48]. Keeping the saturation magnetization constant and the controlled changing of a hard magnet into soft magnet and vice versa could aid such material for futuristic applications in magnetic devices. Moreover, precipitating metastable nanocrystalline phases in amorphous metallic alloys allows for control of the magnetic properties of soft magnetic materials.

6.3 STRUCTURE

Preferably, the characterization of metallic glasses is achieved by the existence of a bulge structure in the X-ray diffraction test. The obtained patterns are useful in obtaining the atomic coordination and partial distribution functions; however, for silicate glasses, no description is present based on random networks. Moreover, for the description of metallic glass, the model which has been found to be most favorable is the polytetrahedral structure. There is an inevitable rise of other configurations owing to the drawback of tetrahedra in the packing space. Figure 6.10 shows the schematic representation of quantity used for calculation of pileup parameter.

6.3.1 CRYSTALLIZATION

Glasses are metastable in nature and, upon heating, devitrification takes place. Three broad categories are present for crystallization reactions, namely, eutectic, primary, and polymorphous crystallization. In eutectic crystallization, according to the eutectic reaction, the glassy matrix separates into two crystalline phases. Tiwari et al. [49]

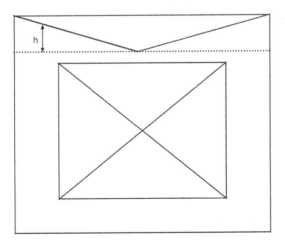

FIGURE 6.10 Schematic representation of quantity used for calculation of pileup parameter.

studied the dependence of eutectic crystallization on annealing temperatures. In primary crystallization, precipitation of the same composition as that of the matrix occurs. However, precipitation of the same composition as that of the matrix occurs. When the temperature is below the glass transition temperature, interlamellar spacing increases and the reverse behavior is observed for temperatures above the glass transition temperature. Growth behavior and nucleation of zirconium-based metallic glass have been reported by Banerjee et al., along with the distinct microstructure. A remarkable characteristic in various BMGs is their ability to form nanocrystals. Phase separation is the cause of the crystal growth phenomenon in the glass or the liquid [50]. Determination of the scale of crystallization takes place due to such fine-scale separation preceding crystallization. A variety of strategies and an atom probe had been utilized by Loffler et al. [51] to support this concept. However, slow growth and high nucleation rate could be lead by large number of quenched-in nuclei. Moreover, by the addition of elements, such as Ag, Pd, Au, and Pt, nanocrystallization in Zr alloys is prompted. Such processes lead to clustering in glass and possess a strong interaction with Zr. Moreover, a comparison between Fe_{40}-Ni_{40}-P_{14}-B_6, which is slowly cooled, and crystallization of rapidly solidified thin ribbons are also of particular interest [49,52].

Table 6.2 gives a comparison between parameters relating to crystallization. The gas formation in solid state is led by the difference between Tg and Tx, defined by the extent of the supercooled region by 42 K. A constant nucleation growth is followed by the crystallization of bulk glasses. Due to the presence of ribbons which may have had heterogeneous impurities, the nucleation rate in the bulk glass is of four orders of magnitude. Moreover, a difference in the size of the crystal in both the cases is present by an order of magnitude.

TABLE 6.2
Comparison of Crystallization Parameters of Bulky Glassy $Fe_{40}Ni_{40}P_{14}B_6$ and Ribbons

Parameter	Melt Spun	Bulk Metallic Glass
Cooling rate (K/s)	10^5–10^6	10^2–10^3
Shape	Ribbons	Spheres
Thickness	50 μm	2 mm
Tg (°C)	387	367
Tx (°C)	393	409
Avrami exponent	3.96	4.05
Nucleation rate (nuclei/m³-s)	~10^{15}	~10^{11}
Growth rate (m/s)	~10^{-9}	~10^{-7}
Microstructural scale (μm)	~3	~30

Source: Tiwari, R., et al., Z. Metallkde, 78, 275–279, 1987; Shen, T.D., and Schwarz, R.B., ActaMaterialia, 49, 837–847, 2001; Ranganathan, S., and Von Heimendahl, M., J. Mater. Sci., 16, 2401–2404, 1981.

6.3.2 Classification

According to Siegel [54], nanocomposite (NC) materials are classified as follows: (i) one-dimensional (1D) filamentary structures, (ii) two-dimensional (2D) layered or lamellar structures, and (iii) three-dimensional (3D) equiaxed grain nanostructured materials. The structure of a NC material is made of two segments: crystallites and interfacial parts. The crystallites have distinctive crystallographic arrangements, while the interfacial parts make the intercrystalline organize, which differs locally. A NC material is described by a high measure of interfaces, where particles are situated at the interface. The most essential parameter of nanocrystalline single-stage alloy metals is the grain estimate. In this way, it is of extreme significance to precisely determine the grain measure. Multiphase nanocrystalline combinations can indicate diverse morphologies. For example, a microstructure of equiaxed grains of various stages can be framed by crystallization of anamorphous compound [56] or by mechanical alloying [56], and a eutectic microstructure with a lamellae width of nanometers [57]. The structure of the crystallites and the nuclear structure at the grain limits in NC materials is equivalent in their coarse-grained partners. Siegel [54] observed this dependent phenomena on confirmations from XRD, X-ray, Mössbauer spectroscopy, Raman spectroscopy, HRTEM, and PC. The reorientation of the nano-sized limits is additionally improved in correlation [58]. HRTEM demonstrates that NC materials comprise little crystallites of various crystallographic introduction isolated by grain limits despite the fact that the grains contain crystalline deformities, for example, separations, twin limits, and stacking deficiencies. It has additionally been noticed that NC materials have vast mean square grid strains, which increases with diminishing grain measure. This strain seems to rely upon the thermal and mechanical properties.

6.4 PROPERTIES OF METALLIC GLASSES (MGs), MG-COMPOSITES, AND NANOCOMPOSITE (NC) MATERIALS

The mechanical properties of materials need to be identified before their practical applications. Therefore, many studies have been conducted in this field. The present section gives an examination of the mechanical conduct of MGs, MG nanocomposite materials, and their crystalline partners.

6.4.1 MGs

As previously mentioned, the characterizing qualities of MGs are their absence of crystallinity, and the related absence of microstructural highlights, for example, grains and stage limits. These properties impact the remarkable mechanical properties of MGs. They show high quality [59], moderately high strain and pressure [60], and their modulus can be 30% lower than the crystalline partners. Additionally, the strain limit of MGs can even surpass 2% before failure (under ductile or compressive conditions). This value is much higher (multiple times) than that found in regular metallic composites. In any case, MGs will, in general, fail after subsequent yielding with constrained plastic deformation.

6.4.2 Deformation Mechanisms

Two fundamental methods of distortion can be recognized in metallic glasses: homogeneous and inhomogeneous stream. The homogeneous flow occurs at low stresses and high temperatures (T > 0.7 Tg). This procedure is identified with viscous flow, where strain appears to be dispersed continuously between various volume components inside the material. Every volume component of the example adds to the strain [61]. On the contrary, at temperatures well underneath the glass change (T < 0.7 Tg) and at high anxieties, MGs experience inhomogeneous deformation by concentrating plastic strain into nanoscale shear groups [62]. It is observed that MGs display the phenomenon called "strain softening." The strain expansion makes the material mild, which leads to deformation of material under low strain rate. Shear band arrangement or shear limitation in MGs has been viewed as an immediate outcome of strain softening. Interestingly, crystalline materials experience "strain solidifying." The material becomes hard and, with the expansion in strain, the material becomes harder and consequently harder to additionally distort it. Some theoretical models have been advanced to clarify the inhomogeneous disfigurement in MGs. It is broadly realized that the shear banding of MGs happens as a result of developments and self-associations flow events. These phenomena can be attributed to a progression of discrete nuclear hops in the glass, at locales where the overabundance of free volume is sufficiently huge. Moreover, this procedure happens by means of diffusion [63]. A second speculation recommends that local adiabatic heating happens in shear groups, which prompts a decline in viscosity [64,65]. This adiabatic heating could prompt an increment in the temperature to a dimension above Tg or even to the liquefying purpose of the alloy. Experimental evidence is accessible for both the expansion in free volume and rise in temperature in shear bands during deformation.

6.4.3 MG Composites

In metallic glass composites, a second phase is presented in amorphous matrix with two fundamental objectives: advancing shear band initiation and spreading (to convey the plastic strain all the more extensively to maintain a distance from the catastrophic confinement) and hindering shear band propagation. Since the principle highlights of MG composites have been clarified in the past area (Section 6.3.2), just a few models are given here. The second stage can exist in numerous types, shapes (particles or dendrites), and diverse sizes (from nuclear groups to micrometer-sized). For example, the size can be controlled by modifying the solidification parameters (e.g., cooling rate). Furthermore, the plasticity can be expanded when nanocrystals or stage division is produced. Deformation induced nanocystallization in which an optional stage is accelerated before distortion inside the underlying glassy stage. However, crystallization prompted during a pressure test was viewed as connected with the generally low stability of the supercooled fluid stage and the high change rate of the essential precipitation stage. Other works suggest that the starting point of plasticity is firmly identified with the SRO/MRO existing in the glassy phase [66], and different authors credited the expansive versatility to the in-situ nanocrystallization produced by plastic flow within the shear bands [67].

6.4.4 NC MATERIALS

NC materials show predominant properties and ordinary coarse-grained materials. This is due to the simple fine-grain sizes and thus high thickness of interfaces in NCs. The grain measure has a critical effect on the yield quality. The connection between yield pressure, σ_y, and grain estimate d is given by the Hall-Petch condition [68]:

$$\sigma_y = \sigma_o + kd^{-\frac{1}{2}} \tag{6.2}$$

where σ_0 is the cross-section area and k is a material ward steady. Yield quality of the greater part of composites increments with diminishing the grain estimate. However, the decline of flow stress with diminishing grain estimate beneath 10 nm, first identified by Chokshi et al. [68] in the nanocrystalline routine, was likewise found and confirmed by atomic elements computations in face-centered cubic (FCC) metals. This phenomenon is called inverse Hall-Petch response, and it is attributed to the developing job of grain-boundary sliding. In grain-boundary reinforcing, the grain limits acts as pinning points, hindering further dislocation propagation. Since the cross-section structure of contiguous grains varies in orientation, it requires more energy for a separation to alter direction and move into the neighboring grain. The grain limit is likewise considerably more disarranged than the inner parts of the grains, which additionally keeps the separations from moving in a persistent slip plane. Obstructing this separation development will hinder the beginning of plasticity and subsequently will build the yield quality of the material. It is notable that grain measure has a solid impact on yield quality as well as on ductility and durability of grain measure (>1 μm) materials. However, some NC materials indicate diminished malleability and, in some cases, a fragile nature. This is apparently because of the inability of dislocation generation and restricted movement at these nanometer grain sizes. Moreover, nanocrystalline metals are characterized by a low solidifying rate, which is an immediate result of the low thickness of dislocations experienced after plastic deformation. This low solidifying rate leads malleability and low elastic flexibility. Moreover, NC materials can demonstrate malleability at room temperature if the plastic deformation did not depend on the dislocation generation and movement, e.g., by deformation induced by grain limit sliding. Besides, it has additionally been demonstrated that the malleability of a NC material is enhanced by expanding the thickness of development twins by annealing [69].

Different systems have been proposed to represent the plastic deformation in NC materials. The most significant are described below [70]:

1. *Pile-up break down:* As the grain measure is diminished, the quantity of dislocations heaped up against grain limit in general continuously increment. In the meantime, an increased stress level is expected to create a similar number of dislocations at the heap-up to accommodate plastic flow. However, at the critical grain estimate, the idea of heap to clarify the plastic stream can no longer be implemented and elective twisting instruments are enacted. This is identified with the converse Hall-Petch regime.

2. *Grain-boundary sliding:* Under shear pressure, one layer of grains slides regarding the other, creating a shear strain simultaneously. Plastic deformation occurs by virtue of the top layer of grains translating to one direction.

3. *Core and mantle model:* The model depends on dislocation generation at neighboring grain limits and on the development of work-solidified grain limit layers. The presence of grain-limit edges with a separation of 10–100 nm gives a supply of nucleation locales of dislocations that, upon discharge into the grains, cross-slip and makes a solidified layer near the grain limits. As the grain estimate is decreased, the proportion between volume divisions of the mantle and center increments, giving an expansion in yield pressure. Past a basic grain estimate, the edges can never again work and don't advance plastic twisting in the nanocrystalline regime.

4. *Shear band formation:* For grain sizes less than d < 300 nm, shear band improvement is regularly observed, following the beginning of plastic disfigurement. This has been related to changes in strain solidifying conduct at those grain sizes since the capacity to work solidify by the expansion in separation thickness is lost.

5. *Mechanical twinning:* Mechanical twinning (occurrence of planar defects) and slip (linear defects) can be considered as competing forms. In FCC metals and alloys, the twinning stress is identified with the stacking fault energy. The decline in grain measure is relied upon to render twinning progressively troublesome. In any event, this is the thing that ordinary materials science predicts in the micrometer regime [71]. In spite of the fact that components for twinning in nanocrystalline metals stay unclear, some preliminary assessments reveal that (a) the conventional nucleation system separates in the nanoscale; (b) there are local stress concentrations (triple points, etc.) that raise the pressure significantly; (c) the partial dislocation increments at the nanoscale area, supporting twinning.

REFERENCES

1. Jun. W.K., Willens, R.H., Duwez, P.O.L. (1960), "Non-crystalline structure in solidified gold–silicon alloys," *Nature*, Vol. 187, pp. 869–870.
2. Telford, M. (2004), "The case for bulk metallic glass," *Materials Today*, Vol. 7, pp. 36–43.
3. Liu, Y.H., Wang, G., Wang, R.J., Zhao, D.Q., Pan, M.X., Wang, W.H. (2007), "Super plastic bulk metallic glasses at room temperature," *Science*, Vol. 315, pp. 1385–1388.
4. Guo, H., Yan, P.F., Wang, Y.B., Tan, J., Zhang, Z.F., Sui, M.L., Ma, E. (2007), "Tensile ductility and necking of metallic glass," *Nature Materials*, Vol. 6, pp. 735–739.
5. Inoue, A. (2000), "Stabilization of metallic supercooled liquid and bulk amorphous alloys," *ActaMaterialia*, Vol. 48, pp. 279–306.
6. Inoue, A. and Takeuchi, A. (2011), "Recent development and application products of bulk glassy alloys," *ActaMaterialia*, Vol. 59, pp. 2243–2267.
7. Schuh, C.A., Hufnagel, T.C., Ramamurty, U. (2007), "Mechanical behaviour of amorphous alloys," *ActaMaterialia*, Vol. 55, pp. 4067–4109.
8. Lowhaphandu, P. and Lewandowski, J. (1998), "Fracture toughness and notched toughness of bulk amorphous alloy: Zr-Ti-Ni-Cu-Be," *ScriptaMaterialia*, Vol. 38(12). doi:10.1016/S1359-6462(98)00102-X.

9. Hofmann, D.C. (2013), "Bulk metallic glasses and their composites: A brief history of diverging fields," *Journal of Materials*, Vol. 2013, doi:10.1155/2013/517904.
10. Demetriou, M.D., Kaltenboeck, G., Suh, J.Y., Garrett, G., Floyd, M., Crewdson, C., Hofmann, D.C. et al. (2009), "Glassy steel optimized for glass-forming ability and toughness," *Applied Physics Letters*, Vol. 95, p. 041907.
11. Conner, R.D., Johnson, W.L., Paton, N.E., Nix, W. (2003), "Shear bands and cracking of metallic glass plates in bending," *Journal of Applied Physics*, Vol. 94, pp. 904–911.
12. Conner, R.D., Li, Y., Nix, W.D., Johnson, W. (2004), "Shear band spacing under bending of Zr-based metallic glass plates," *ActaMaterialia*, Vol. 52, pp. 2429–2434.
13. Demetriou, M.D., Launey, M.E., Garrett, G., Schramm, J.P., Hofmann, D.C., Johnson, W.L., Ritchie, R.O. (2011), "A damage-tolerant glass," *Nature Materials*, Vol. 10, pp. 123–128.
14. Hofmann, D.C. and Johnson, W.L. (2010), "Improving ductility in nanostructured materials and metallic glasses: Three laws," *Materials Science Forum*, Vol. 633–634, pp. 657–663.
15. Singh, D., Mandal, R.K.,Tiwari, R.S. and Srivastava, O.N. (2016) "Mechanical behaviour of Zr-based metallic glasses and their nanocomposites," *Metallic Glasses – Formation and Properties* (Ed.), Movahedi, B., IntechOpen, doi:10.5772/64221.
16. Basu, J. and Ranganathan, S. (2003), "Bulk metallic glasses: A new class of engineering materials," *Sadhana: Academy Proceedings in Engineering Sciences*, Vol. 28, pp. 783–798.
17. Soubeyroux, J.L., Puech, S., Donnadieu, P., Blandin, J.J. (2007), "Synthesis and mechanical behaviour of nanocomposite Mg-based bulk metallic glasses," *Journal of Alloys and Compounds*, Vol. 434–435, pp. 84–87.
18. Fornell, J., González, S., Rossinyol, E., Suriñach, S., Baró, M.D., Louzguine-Luzgin, D.V., Perepezko, J.H., Sort, J., Inoue, A. (2010), "Enhanced mechanical properties due to structural changes induced by devitrification in Fe–Co–B–Si–Nb bulk metallic glass," *ActaMaterialia*, Vol. 58, pp. 6256–6266.
19. Mayer, C., Chevalier, B., Gorsse, S. (2010), "Magnetic and magnetocaloric properties of the ternary Gd-based metallic glasses $Gd_{60}Mn_{30 \times 10}$, with X = Al, Ga, In," *Journal of Alloys and Compounds*, Vol. 507, pp. 370–375.
20. Eckert, J., Das, J., Pauly, S., Duhamel, C. (2007), "Mechanical properties of bulk metallic glasses and composites," *Journal of Materials Research*, Vol. 22, pp. 285–301.
21. Choi-Yim, H. and Johnson, W.L. (1997), "Bulk metallic glass matrix composites," *Applied Physics Letters*, Vol. 71, pp. 3808–3810.
22. Szuecs, F., Kim, C.P., Johnson, W.L. (2001), "Ductile phase reinforced bulk metallic glass composite," *ActaMaterialia*, Vol. 49, pp. 1507–1513.
23. Kokotin, V., Hermann, H., Eckert, J. (2012), "Theoretical approach to local and effective properties of BMG based matrix-inclusion nanocomposites," *Intermetallics*, Vol. 30, pp. 40–47.
24. Roy, B., Roy, S., Chakravorty, D. (1994), "Electrical properties of glass-metal nanocomposites synthesized by electrodeposition and ion exchange/reduction techniques," *Journal of Materials Research*, Vol. 9, pp. 2677–2687.
25. Rossinyol, E. and Surin, S. (2010), "Enhanced mechanical properties due to structural changes induced by devitrification in Fe–Co–B–Si–Nb bulk metallic glass," *ActaMaterialia*, Vol. 58, pp. 6256–6266.
26. Albe, K., Ritter, Y., Şopu, D. (2013), "Enhancing the plasticity of metallic glasses: Shear band formation, nanocomposites and nanoglasses investigated by molecular dynamics simulations," *Mechanics of Materials*, Vol. 67, pp. 94–103.
27. Ashby, M.F. and Greer, A.L (2006), "Metallic glasses as structural materials," *ScriptaMaterialia*, Vol. 54, pp. 321–326.
28. Misra, D.K., Sohn, S.W., Kim, W.T., Kim, D.H. (2009), "Plastic deformation in nanostructured bulk glass composites during nanoindentation," *Intermetallics*, Vol. 17, pp. 11–16.

29. Eckert, J., Kübler, A., Schultz, L. (1999), "Mechanically alloyed $Zr_{55}Al_{10}Cu_{30}Ni_5$ metallic glass composites containing nanocrystalline W particles," *Journal of Applied Physics*, Vol. 85, pp. 7112–7119.

30. Calin, M., Eckert, J., Schultz, L. (2003), "Improved mechanical behaviour of Cu–Ti-based bulk metallic glass by in-situ formation of nanoscale precipitates," *ScriptaMaterialia*, Vol. 48, pp. 653–658.

31. Branagan, D.J., Tang, Y.L., Sergueeva, A.V, Mukherjee, A.K. (2003), "Low temperature superplasticity in a nanocomposite iron alloy derived from a metallic glass," *Nanotechnology*, Vol. 14, pp. 1–15.

32. Bahrami, F., Amini, R., Taghvaei, A.H. (2017), "Microstructure and corrosion behavior of electrodeposited Ni-based nanocomposite coatings reinforced with $Ni_{60}Cr_{10}Ta_{10}P_{16}B_4$ metallic glass particles," *Journal of Alloys and Compounds*, Vol. 714, pp. 530–536.

33. Gorsse, S., Chevalier, B., Orveillon, G. (2008), "Magnetocaloric effect and refrigeration capacity in $Gd_{60}Al_{10}Mn_{30}$ nanocomposite," *Applied Physics Letters*, Vol. 92, p. 122501.

34. Branagan, D.J., Swank, W.D., Haggard, D.C., Fincke, J.R. (2001), "Wear-resistant amorphous and nanocomposite steel coatings," *Metallurgical and Materials Transactions A*, Vol. 32, pp. 2615–2621.

35. Zhu, T.J., Yan, F., Zhao, X.B., Zhang, S.N., Chen, Y., Yang, S.H. (2007), "Preparation and thermoelectric properties of bulk in-situ nanocomposites with amorphous/nanocrystal hybrid structure," *Journal of Physics D: Applied Physics*, Vol. 40, p. 6094.

36. Johnson, W.L. (2002), "Bulk amorphous metal: An emerging engineering material," *The Journal of the Minerals, Metals & Materials Society (TMS)*, Vol. 54, pp. 40–43.

37. Angell, C.A. (1995), "Formation of glasses from liquids and biopolymers," *Science*, Vol. 267, pp. 1924–1935.

38. Kim, J.J., Choi, Y., Suresh, S., Argon, A.S. (2002), "Nanocrystallization during nanoindentation of a bulk amorphous metal alloy at room temperature," *Science*, Vol. 295, pp. 654–657.

39. Raja, V. and Ranganathan, S. (1988), "Effect of molybdenum and silicon on the electrochemical corrosion behaviour of FeNiB metallic glasses," *Corrosion*, Vol. 44, pp. 263–270.

40. McHenry, M.E. and Laughlin, D.E. (2000), "Nano-scale materials development for future magnetic applications," *ActaMaterialia*, Vol. 48, pp. 223–238.

41. Hsu, C., Hsieh, M., Fu, C., Huang, Y. (2015), "Effects of multicore structure on magnetic losses and magnetomechanical vibration at high frequencies," *IEEE Transactions on. Magnetics*, Vol. 51, pp. 1–4.

42. Das, S., Choudhary, K., Chernatynskiy, A., Yim, H.C., Bandyopadhyay, A.K., Mukherjee, S. (2016), "Spin-exchange interaction between transition metals and metalloids in soft-ferromagnetic metallic glasses," *Journal of Physics: Condensed matter*, Vol. 28, p. 216003.

43. Wang, J., Li, R., Hua, N., Zhang, T. (2011), "Co-based ternary bulk metallic glasses with ultrahigh strength and plasticity," *Journal of Materials Research*, Vol. 26, pp. 2072–2079.

44. Coey, J.M.D. and Sun, H. (1990), "Improved magnetic properties by treatment of iron-based rare earth intermetallic compounds in ammonia," *Journal of Magnetism and Magnetic Materials*, Vol. 87, pp. L251–L254.

45. Fuerst, C.D. and Brewer, E.G. (1993), "High-remanence rapidly solidified Nd-Fe-B: Die-upset magnets," *Journal of Applied Physics*, Vol. 73, pp. 5751–5756.

46. Blázquez, J.S., Franco, V., Conde, A., Kiss, L.F. (2003), "Soft magnetic properties of high-temperature nanocrystalline alloys: Permeability and magnetoimpedance," *Journal of Applied Physics*, Vol. 93, pp. 2172–2177.

47. Wang, W.H., Pan, M.X., Zhao, D.Q., Hu, Y., Bai, H.Y. (2004), "Enhancement of the soft magnetic properties of FeCoZrMoWB bulk metallic glass by microalloying," *Journal of Physics: Condensed Matter*, Vol. 16, p. 3719.

48. Lachowicz, H.K., Zaveta, K., Slawska-Waniewska, A. (2002), "Magnetic properties of partially devitrified metallic glasses," *IEEE International Magnetics Conference (INTERMAG)*, Amsterdam, the Netherlands, doi:10.1109/INTMAG.2002.1001007.

49. Tiwari, R., Heimendahl, M., Ranganathan, S. (1987), "On the variation of interlamellar spacing in the crystallization product of the amorphous alloy $Fe_{40}Ni_{40}P_{14}B_6$," *Z. Metallkde*, Vol. 78, pp. 275–279.

50. Busch, R., Schneider, S., Peker, A., Johnson, W.L. (1995), "Decomposition and primary crystallization in undercooled $Zr_{41.2}Ti_{13.8}Cu_{12.5}Ni_{10.0}Be_{22.5}$melts," *Applied Physics Letters*, Vol. 67, p. 1544.

51. Löffler, F., Johnson, J., Wagner, W., Thiyagarajan, P. (2000), "Comparison of the decomposition and crystallization behaviour of Zr and Pd-based bulk amorphous alloys," *Materials Science Forum*, Vol. 343–346, pp. 179–184.

52. Shen, T.D. and Schwarz, R.B. (2001), "Bulk ferromagnetic glasses in the Fe–Ni–P–B System," *ActaMaterialia*, Vol. 49, pp. 837–847.

53. Ranganathan, S. and Von, Heimendahl, M. (1981), "The three activation energies with isothermal transformations: Applications to metallic glasses," *Journal of Materials Science*, Vol. 16, pp. 2401–2404.

54. Siegel, R.W. (1993), "Nanostructured materials-mind over matter," *Nanostructured Materials*, Vol. 4, pp. 121–138.

55. Gleiter, H. (1989), "Nanocrystalline materials," *Progress in Materials Science*, Vol. 33, pp. 223–315.

56. Lu, K. (1996), "Nano-crystalline metals crystallized from amorphous solids: Nanocrystallization, structure, and properties," *Materials Science and Engineering R: Reports*, Vol. 16, pp. 161–221.

57. He, G., Eckert, J., Löser, W., Schultz, L. (2002), "Novel Ti-base nanostructure–dendrite composite with enhanced plasticity," *Nature Materials*, Vol. 2, pp. 33–37.

58. Meyers, M.A., Mishra, A., Benson, D.J. (2006), "Mechanical properties of nanocrystalline materials," *Progress in Materials Science*, Vol. 51, pp. 427–556.

59. Inoue, A., Shen, B., Koshiba, H., Kato, H., Yavari, A.R. (2003), "Cobalt-based bulk glassy alloy with ultrahigh strength and soft magnetic properties," *Nature Materials*, Vol. 2, pp. 661–663.

60. Greer, A.L. (1995), "Metallic glasses," *Science*, Vol. 267, pp. 1947–1953.

61. Chen, H.S. and Goldstein, M. (1972), "Anomalous viscoelastic behaviour of metallic glasses of Pd–Si-based alloys," *Journal of Applied Physics*, Vol. 43, pp. 1642–1648.

62. Argon, A.S (1979), "Plastic deformation in metallic glasses," *ActaMetallurgica*, Vol. 27, pp. 47–58.

63. Spaepen, F. (1977), "A microscopic mechanism for steady state inhomogeneous flow in metallic glasses," *ActaMetallurgica*, Vol. 25, pp. 407–415.

64. Dai, L.H., Yan, M., Liu, L.F., Bai, Y.L (2005), "Adiabatic shear banding instability in bulk metallic glasses," *Applied Physics Letters*, Vol. 87, p. 141916.

65. Leamy, H.J., Wang, T.T., Chen, H.S (1972), "Plastic flow and fracture of metallic glass," *Metallurgical and Materials Transactions*, Vol. 3, p. 699.

66. Lee, S.W., Huh, M., Fleury, E., Lee, J.C. (2006), "Crystallization-induced plasticity of Cu–Zr containing bulk amorphous alloys," *ActaMaterialia*, Vol. 54, pp. 349–355.

67. Chen, M., Inoue, A., Zhang, W., Sakurai, T. (2006), "Extraordinary plasticity of ductile bulk metallic glasses," *Physical Review Letters*, Vol. 96, p. 245502.

68. Chokshi, A.H., Rosen, A., Karch, J., Gleiter, H. (1989), "On the validity of the Hall-Petch relationship in nanocrystalline materials," *ScriptaMetallurgica*, Vol. 23, pp. 1679–1683.

69. Zhu, Y.T. and Liao, X. (2004), "Retaining ductility," *Nature Materials*, Vol. 3, pp. 351–352.
70. Suryanarayana, C. and Koch, C.C (2000), "Nanocrystalline materials–Current research and future directions," *Hyperfine Interactions*, Vol. 130, p. 5.
71. Tlili, A., Pailhès, S., Debord, R., Ruta, B., Gravier, S., Blandin, J.J., Blanchard, N., Gomès, S., Assy, A., Tanguy, A., Giordano, V.M. (2017), Thermal transport properties in amorphous/nanocrystalline metallic composites: A microscopic insight. *ActaMaterialia*, Vol. 136, pp. 425–435.

Index

Note: Page numbers in italic and bold refer to figures and tables, respectively.